BIOSAFETY AND BIOETHICS

BIOSAFETY AND BIOETHICS

Editor
Rajmohan Joshi

ISBN : 978-81-8205-377-9

Published, 2018

Published by

Isha Books
D-69, New Gupta Colony,
Near DDA Market
Delhi — 110009
Ph.: +91-9212142040

Preface

The recent advances in the field of biotechnology have brought into focus several ethical and safety issues. The inventions in the field of genetic engineering and related fields of molecular biology will affect not only ourselves but the plants, microorganisms, animals and the entire environment and the way we practise agriculture, medicine and food processing. An increase in our ability to change life forms in recent years has given rise to the new science of 'bioethics'. Bioethics can be defined as the systematic study of human conduct in the area of life sciences and health sciences, keeping in view the moral values and principles existing during the particular period.

During the last few decades there has been a growing feeling that technology has led to many problems along with benefits leading to a strong anti-technology trend. Adverse opinion regarding science, even if it is in a small minority of people, can lead to protest actions as seen in recent examples of protests over the release of genetically modified organisms, animal rights movement, sex determination of foetuses and abortion etc. There have been terror campaigns conducted against scientists, and destruction of laboratories and test farms concerned with genetically modified crops. This is true of several countries and the anti-biotechnology movement is gathering wind. There has been a feeling that the scientists are not transparent and on the contrary are arrogant about their research capabilities.

The concerns regarding the introduction of GMOs is that the new genes may enter other organisms or that the new organisms themselves may replace existing organisms in the ecosystem. The ecological system is very complex and we cannot yet predict these results, but we must walk carefully as the environment is already in bad shape because of over-exploitation. There is concern about protecting species diversity. There is also concern regarding the powers to control nature being vested in private companies where commercial concerns dominate. While anti-biotechnology activists are over rating the risks of biotechnology, it is time for the scientists to make a scientific and objective analysis of the social issues involved, and make it known to the public who will, otherwise, be carried away by the emotional rhetoric by the less informed but highly vocal section of the society.

The present book discusses the biosafety and bioethical issues the modern society confronts. Topics such as biotech development, impact of biotechnology on biosafety, biotech products and ethical issues, governance of biosafety, environmentally responsible use of biotechnology, etc., are describe in detail. This book is destined to become an essential reading for students, teachers and professionals in all fields of life sciences.

Editor

Contents

Preface v

1. Modern Biotechnology and Biosafety 1
2. Biotechnology and Issues of Biosafety 11
3. Biotechnology Development in India 41
4. National Biotechnology Development Strategy 59
5. Implications of Biotechnology in Trade Communities 119
6. Biotechnological Products and Ethical Issues 148
7. Governing Biosafety in India 182
8. Environmentally Responsible Use of Biotechnology 214
9. Cartagena Protocol on Biosafety to the Convention on Biological Diversity 231

Bibliography 269

Index 271

1

Modern Biotechnology and Biosafety

Biodiversity — the variety of life and of habitats on Earth — is vital to human welfare. Biodiversity loss also has social consequences, in its impact on people's livelihoods and lifestyles — this is the unmeasured cost of losing cultural traditions. The loss of key elements of an ecosystem can alter the balance between its components and lead to long-term or permanent changes. Biodiversity loss can also affect human health, as our health is largely dependant on the quality of the ecosystem in which we live — loss of plant or animal species can affect the quality of water or soil, for instance.

The main pressures on biodiversity result from land use changes (usually associated with increasing populations); unsustainable use and exploitation of natural resources (especially fisheries, agriculture, and forestry); global climate change; and industrial pollution. At the same time, biotechnology is introducing new organisms and their effect on existing organisms and habitats also needs to be considered. In some instances, these pressures can actually be positive for biodiversity.

Agricultural activity sometimes improves the habitat and even helps increase the variety of species; the Mediterranean basin is considered a biodiversity "hot spot" in part because of its human-induced agricultural biodiversity. However, the available evidence suggests that, in most regions of the world, the effects of economic activity are negative for biodiversity. This Policy Brief looks at what governments can do to place a value on and help preserve biodiversity and ensure biosafety.

Biodiversity in general is more dispersed and less easily targeted for collective policy. There is also less room for error — if air becomes more polluted, steps can be taken to reduce the pollution, but if an ecosystem is compromised by human encroachment, its loss may be permanent. Given that our knowledge of the complexity and inter-linkages of ecosystems are only cursory, loss of biodiversity may be more important than any immediate gains. Before policy makers take decisions about whether to risk destroying a habitat for economic gain (such as a housing development), they need to be able to set a relative value for those gains and for the biodiversity at risk.

But most environmental amenities or biodiversity — landscapes, species of insect, forest habitats — are not bought or sold and therefore a price cannot be set on their use without some form of policy intervention. Without collective action, over-exploitation of such common assets is the likely result. In the case of biodiversity, at the extreme this leads to such destructive practices as "slash and burn" farming on fragile soils. Overall, there is a good case to be made for vigorous policy intervention to help preserve both the extent and the quality of biodiversity.

Once governments decide to act to protect biodiversity, they need to choose policies that reflect the relative value people place on it. Placing a market value on biodiversity may sound difficult, but the strength of people's desire to maintain biodiversity can be tested at many levels. For example, evidence can be gleaned from people's willingness to visit natural areas and the pleasure they derive from nature hikes and other non-destructive outdoor activities.

The rapid growth of the global eco-tourism industry also demonstrates a strong willingness by people to pay substantial amounts of money to see natural phenomena that are not available locally.

Many studies have been conducted to quantify the value people place on biodiversityrelated goods and services relative to other economic outputs. Governments can use this information to craft public policy that will encourage the use and conservation of biodiversity in a way that reflects its relative value. OECD work on the economic aspects of biodiversity explores the types of policies and instruments that will lead to uses of biodiversity-related resources that best reflect relative preferences. Part of this work seeks to ensure that the biodiversity-related impacts of economic activity are an explicit part of the day-to-day decisions that consumers and producers face.

For example, imposing a higher cost on products using virgin resources than those using recycled resources could be used to reflect impacts on biodiversity. When biodiversity policy accounts for both private and public values of biodiversity, as well as for the consequences on all affected individuals (including future generations), the use of biodiversity resources will be consistent with achieving the greatest net benefit to society over the long term. The choice of policy instruments is complex and depends on specific institutional, economic and social needs.

Policy options should be systematically analysed to minimise the costs of public administration, monitoring and enforcement, as well as the private costs of implementation. Since market-based instruments that change the prices of biodiversity-impacting products (e.g. wood products) are often cost-effective — and generally under-utilised — they should be promoted. However, it will still often be necessary to use non-market-based instruments (e.g. regulatory initiatives such as banning trade in endangered species) in the policy mix. There is also a need to work at the international level to implement biodiversity management policies, for example in development co-operation or in policies to protect migratory species and aquatic resources. Moreover, biodiversity-related resources have non-use values to people everywhere that need to be reflected in local use decisions - without compromising local economic development.

Incentive measures are the basis of a market approach to biodiversity management. Such measures help reconcile differences between the value of biodiversity-related resources to individuals and the value of biodiversity to society as a whole. They increase the cost of activities that damage ecosystems important for biodiversity and reward biodiversity conservation and enhancement/restoration. Farmers who receive a government payment for maintaining biological diversity on their land, for example, will be more willing to use farm practices that sustain biodiversity values. The idea of using incentives to achieve particular biodiversity objectives often requires a method for gauging the value of biodiversity.

Unfortunately, reaching consensus on such method(s) is not always easy. Economic valuation can sometimes offer a solution if monetary measures of the impacts involved can be obtained or implied — such as the cost of travel, and related expenses, for visiting natural areas. However, it is more difficult to quantify the more aesthetic or cultural values involved, for example, in non-use of biodiversity, such as declaring a mountainside an ecological conservation area. For those non-use values, techniques to get people to "reveal" biodiversity's value to them are needed.

Markets are created by removing barriers to trade, including the establishment and assignment of well-defined and stable property and/or user rights. Market creation is based on the premise that holders of these rights will maximise the value of their resources over time, thereby optimising biodiversity use, conservation, and restoration.

Market creation therefore involves a broader approach than the simple use of market incentives. Governments have two important roles to play in supporting markets for biodiversity-related resources. First, they need to establish the right framework conditions for private and public operators to supply biodiversityrelated resources efficiently to users. They also need to apply the right policy instruments to ensure that public biodiversity-related goods and services are provided in the most efficient and effective manner.

Three issues need special attention. First, the absence of appropriate information can inhibit the development and implementation of market approaches to biodiversity conservation, use, and restoration. Information can be provided through such mechanisms as labelling, certification and technical capacity-building. Scientific knowledge is also important, so governments need to develop policies that establish the right conditions for new knowledge to emerge related to biodiversity conservation. Indicators to monitor biodiversity change will also be important, along with the active and early engagement of stakeholders in developing and implementing biodiversity management policies.

Local community networks that identify and support local biodiversity objectives can make important contributions in this regard. Also, markets need to be periodically monitored to ensure they actually result in net benefits for society as a whole. A number of specific markets have already developed around biodiversityrelated activities. Examples include: organic agriculture; sustainable forestry; non-timber forest products; genetic resources; and eco-tourism. Two highly successful examples where the instruments themselves created the market are trading in access to fishing rights and transferable development rights to land.

The emergence of private parks in many regions of the world also demonstrates that there is scope for capturing public values in private markets. For those parks, the private value of their uniqueness is high enough to support public biodiversity objectives. However, since the public value of the parks will typically be greater than the private values, economic incentives that capture some of these additional public values would improve the efficiency and effectiveness of biodiversity management.

Importance of Biosafety

Products of modern biotechnology might also affect biodiversity, and thus raise environmental safety issues. Genetically engineered crops (also known as transgenic crops) are no longer an idea of the future, but have well and truly arrived. Crops such as maize, soybean, rapeseed and cotton are being approved for commercial use in an increasing number of countries. The OECD has been active on such issues since the mid-1980s. From 1996 to 2004, there was more than a 47-fold increase in the area grown with transgenic crops worldwide, reaching 81.0 million hectares.

In 2004, there were 14 countries that grew 50,000 hectares or more: the US grew 59% of the world total, followed by Argentina (20%), Canada (6%), Brazil (6%), China (5%), Paraguay (2%), India (1%), and South Africa (1%). In addition, Uruguay, Australia, Romania, Mexico, Spain and the Philippines each had smaller scale cultivation of less than 1% of the total. Most of this cultivation was devoted to soybean (60%), maize (23%), cotton (11%), and rapeseed (6%). So far, most commercialisation has focused on these crops, and the genetic engineering has involved two traits: insect resistance and herbicide tolerance. The trend for an increase in the transgenic crop area seems set to continue, given the large range of genetically engineered crops in research and development

In all OECD countries, a notification or registration has to be made to obtain approval for the commercial use of a genetically engineered crop — whether for planting and growing or for use in human or animal foods. Such approval is always based on a safety assessment by national authorities, which in turn is based on scientific information regarding the crop, its specific trait, and the receiving environment. It is important for governments to ensure that good quality safety information is publicly available and, where possible, to adopt international approaches to risk and safety assessment that ensure the efficiency of the risk assessment process.

One of the primary goals of the OECD's work on biosafety in the past decade has been to promote harmonisation among member countries of notifications and registrations of biotechnology products. Such harmonisation aims to ensure that the information used in risk/safety assessments, as well as the methods used to collect such information, are as similar as possible. This can lead to countries recognising or even accepting information from one another's safety assessments, and it generates significant benefits. It increases mutual understanding among member countries of each other's risk assessments; it avoids duplication of effort; it saves on scarce resources; and it increases the efficiency of the risk/safety assessment process. This in turn improves safety, while reducing unnecessary barriers to trade.

For harmonisation to be possible among member countries, it is important that they have similar approaches to risk/safety assessment. Earlier OECD work on risk assessment demonstrated that such assessment should be based on the characteristics of the organism, the introduced trait, the environment into which the organism is introduced, the interaction between these, and the intended application. This work has formed the basis for environmental risk/safety assessment that is now globally accepted.

The similarity of approach was reinforced by the fact that most genetically engineered organisms are developed from organisms such as crop plants whose biology is well understood. This allows the risk assessor to draw on previous knowledge and experience with the introduction of plants and micro-organisms into the environment. The process takes account of a wide range of attributes including, for example, knowledge and experience with the plant, including its flowering/reproductive characteristics, ecological requirements, and past breeding experiences. It is not just that national authorities use the same concepts and principles in risk assessment.

OECD countries also share a remarkably high degree of similarity in the questions and issues addressed in risk/safety assessments as outlined in national laws, regulations, and guidance documents. Of course, national authorities also request some regulatory information which is specific to the local environment. But nevertheless, much of the information used in risk/safety assessment that relates to the biology of crop plants and micro-organisms is similar or virtually the same in all assessments involving the same organism. So a major focus of the OECD's work on harmonisation is to compile the biological information common to the risk/safety assessment of a number of transgenic products, focusing on two specific categories: the biology of the host species or crop; and traits used in genetic modifications.

The aim is to encourage information-sharing and prevent duplication of effort among countries by avoiding the need to address the same common issues in each application involving the same organism or trait. The resulting Consensus Documents on biology and traits are not intended to be a substitute for a risk/safety assessment, because they address only the generic part of the information that member countries believe is relevant to risk/safety assessment. They are intended to be a "snapshot" of current information, for use during the regulatory assessment of products of biotechnology.

Nevertheless, they make an important contribution to environmental risk/safety assessment and help prevent duplication of effort among countries. An additional challenge is to transfer the knowledge obtained by countries with experience in risk/safety assessment to all those who need it, including non-OECD countries. This enables those involved in risk assessment of transgenic products to easily obtain information developed from previous experiences. The OECD has established information exchange mechanisms and databases to support this process.

One such mechanism, BioTrack Online, has been one of the best sources of information on regulatory developments in OECD countries, as well as field trials and approvals of commercial products. The OECD's Product Database contains information on the safety of genetically engineered crops in commercial use, as well as links to the Web sites of national authorities. BioTrack Online is publicly available on the Internet, and in recent years has been linked with the Biosafety Clearing House (BCH) which is part of the Cartagena Biosafety Protocol to the UN Convention on Biological Diversity. The Cartagena Protocol is intended to lay the foundation for a global system for assessing and managing the impact of living modified organisms on biodiversity.

Genetically Engineered Crops

Confusion can arise when national authorities share information on the same genetically engineered crop if different names or descriptions are used for the same type of maize or cotton. To avoid this problem, the OECD has developed a system of "unique identifiers" for transgenic plants. A unique nine-digit letter and number code is given to each new transgenic plant that is approved for commercial use and becomes its "name" worldwide. So, for instance, a maize developed by Monsanto to be resistant to insect pests has a unique identifier of MON-ØØ810-6, while DD-Ø1951A-7 denotes a cotton developed by DuPont.

The guidance provides for the developers of a new transgenic product to generate the identifier. Once approved, national authorities can forward the unique identifier for inclusion in the OECD's database. OECD countries are already using the system. The OECD has been forwarding unique identifiers to the Biosafety Clearing House which is a key element of the Protocol's activities. The system works well for genetically engineered crops and the OECD is now considering how the identifier tool can be extended beyond crops to micro-organisms and animals. Thanks to unique identifiers, all stakeholders, including the public, will be able to access solid and reliable information when making their judgements about safety.

2

Biotechnology and Issues of Biosafety

History of the Cartagena Biosafety Protocol

The road to the CPB is, roughly speaking, characterized by five phases containing elements typical for any process leading to an international environmental agreement.

Problem Identification

The first phase (1970s and '80s) can be described as the period when the biotechnology issue emerged. As public concern grew over the implications of genetically modified organisms (GMOs) arising from biotechnology, questions arose in various fields. On the scientific front, there were queries regarding the possible harmful effects of GMOs while environmental discussions focused on whether GMOs would further sustainable development or not.

Framework Development

The late 1980s and the beginning of the 1990s saw the development of an international framework to address biosafety issues as well as biosafety guidelines. An Informal Working Group on Biosafety was created in 1985 (comprising UNIDO, UNEP, WHO and later FAO) and the UNIDO Voluntary Code of Conduct for the Release of Organisms into the Environment as well as the OECD Safety Considerations for Biotechnology were issued.

The CBD Negotiation Process

The third phase (1988-1992), partly overlapping phase 2, can be characterised as the Biodiversity Convention (CBD) negotiation process. This process was rather peculiar with regard to the biosafety issue. While biotechnology was at the centre of CBD negotiations from the earliest days, biosafety as an international concern only emerged much later in the process. At the 8th meeting of the CBD process in November 1991, Malaysia tabled a proposal on international safety measures with regard to GMOs. This proposal is the origin of Article 19 (3) of the CBD and therefore also the origin of the CPB.

It contained a core element that prevailed throughout the negotiations and later became a core element of the CPB, although formulated differently. The element was prior informed consent from countries where GMOs are to be introduced. While the issue of an international regulatory mechanism on biosafety was raised barely half a year before concluding the negotiations in order for the CBD to be signed at UNCED in June 1992, it was one of the most difficult to resolve. It is a pity that the negotiation history of the CBD has never been documented. However, it is not too late.

The adoption and signature of the CBD with its Article 19 (3) containing a legal obligation for the Parties to the CBD to consider the need for and modalities of a protocol setting out appropriate procedures, including, in particular, advance informed agreement, in the field of the safe handling and use of any LMO that may have adverse effect on the conservation and sustainable use of biodiversity constituted the first turning point. Article 19 (3) together with Article 8 (g) on domestic measures and some obligations with regard to biosafety contained in Article 19 (4) provided the basis for the fourth phase, i.e. the issue-definition phase.

Issue-Definition Phase

The fourth phase (1992-1995), also partly overlapping the second phase, can be described as the issue-definition phase. It covers the period from the adoption and signature of the CBD until and including the "Jakarta Mandate". As a first step, the then Executive Director of UNEP, Dr. Tolba, established a panel of experts to facilitate the consideration of the need for and modalities of a protocol on biosafety. Only two experts, one from a country and another from an IGO, disagreed with the report. However, the issue of biosafety itself was on the agenda of COP 1, and Parties agreed to establish a process with a view that a decision on whether or not to elaborate a protocol could be taken at COP 2.

In May 1995, a panel of experts (the "Cairo Panel") met in order to prepare a report to be included as a basis for the following step, the meeting of an Open-ended Ad Hoc Group of Experts. However, the report of the Cairo Panel did not contain clear recommendations. The large majority of the Ad Hoc Group of Experts, which met in July 1995 (Madrid), favoured the development of a protocol on biosafety, but there was considerable disagreement on a number of possible elements. Finally, at COP 2 in November 1995 in Jakarta, Indonesia, a decision was taken to establish an Open-ended Ad Hoc Working Group on Biosafety to elaborate a protocol on biosafety specifically focusing on the transboundary movements of LMOs (BSWG). This was the second turning point, which marked the beginning of the actual Protocol negotiations.

It is interesting to note that in the Jakarta Mandate, the notion of "any LMO resulting from biotechnology" was restricted to "LMOs resulting from modern biotechnology". It is also interesting to compare the definition of LMO currently in the CPB with the statement of Malaysia at the Nairobi Final Act Conference (22 May 1992). Malaysia understood the term "living modified organisms" to mean "genetically modified organisms", the notion that a few countries fought so hard - and successfully -to avoid. Probably a number of scientists would argue that the difference between the definition of LMOs in the Protocol and the definition of GMOs is hardly identifiable.

The Negotiation Phase

The final phase (1996-2000) can be defined, broadly speaking, as the negotiation phase, from the first session of the BSWG in July 1996 and to the resumed ExCOP in Montreal in January 2000. This phase is the main subject of the study The Cartagena Protocol on Biosafety. The first sub-phase covers the two first sessions of the BSWG. It constitutes the element-definitions phase resulting inter alia in a decision to invite governments to submit to the third meeting legal texts on most of the substantive elements and to request the Secretariat to develop draft articles on most of the remaining issues. This is the third turning point, agreement to begin building a formula or drafting.

The next sub-phase can be described as the draft-producing and negotiation phase. It includes the remaining four sessions of the BSWG and the ExCOP and constituted two turning points. The fourth turning point, which occurred at the last meeting of the BSWG in February 1999, characterized the final bargaining phase. While delegations had been reluctant to negotiate at the previous meetings, this being the last BSWG, delegations had to take their final positions. Unfortunately, this session failed to produce an agreed protocol text, but the ExCOP, which followed immediately after the last BSWG session to formally adopt the CPB decided to adjourn and revert- to the final bargaining phase, but now on the basis of clean draft protocol, i.e. a draft without brackets.

The third sub-phase under the able and inspiring chairmanship of Juan Mayr resulted in the adoption of the protocol, constituting the sixth turning point, the conclusion of the negotiations. The seventh turning point was the entry into force of the CPB on 11 September 2003. The adoption by the first COP-MOP decisions of crucial importance for the future implementation and effective functioning of the Protocol will no doubt constitute yet another turning point.

Biosafety, Biotechnology and the Environment

The Cartagena Protocol on Biosafety, was adopted in January 2000. While it is generally recognised that biotechnology has potential benefits for the protection of the environment and for improving human health, by the same token questions have been raised about the potential adverse effects of its products on the environment. Over the last thirty years, since the development of recombinant DNA techniques to make specific modifications to organisms for use in medicine, food production and agriculture, biosafety issues have steadily gained recognition at national, regional and international levels. This has been an important factor in the development of the Biosafety Protocol.

The Cartagena Protocol on Biosafety is very important, particularly for developing countries. It is an international agreement that specifically focuses on the transboundary movement of genetically modified organisms (GMOs). Although GMOs are neither inherently risky nor safe, it is generally recognised that the potential to create new genetic combinations and the relatively limited experience with GMOs warrant national and international regulation. Countries always had the sovereign right to regulate GMOs and their products at the national level, and they typically do this by reviewing certain technical information to determine safety. The Protocol now establishes an international, legallybinding framework that allows countries, in particular, those that do not yet have in place a regulatory regime for biosafety, to make informed decisions on the import of GMOs into their country.

Role of Biotechnology in the Protection of the Environment

Positive contributions of biotechnology have been recognized in different areas such as human health, agricultural production of food, feed and fibres and environmental protection. Agricultural biotechnology has received quite substantial attention compared with environmental biotechnology. Nevertheless, the development of bio-engineered crops or transgenic crops has to take into consideration environmental issues when it comes to field-testing and release into the environment. It is estimated that in 2002 Genetically Modified crops covered a total area of 58.1 million hectares, globally, and this total is predicted to increase.

As a broader perspective on the potential benefit of biotechnology in agriculture, it has been argued that better agricultural efficiency will lead to less pressure on agricultural land and consequently less pressure on forests as well as other ecosystems. Further, it is argued that the use of pest- and disease-resistant crops and pesticides from a biological origin does not affect the environment as negatively as the use of persistent, harmful synthetic pesticides. The same can be said for reduction in the use of chemicals, energy and water in industrial processes where (micro)-organisms are used instead of chemicals under high energy input.

In addition to a potential indirect benefit to forests and other ecosystems resulting from less pressure on land use, biotechnology can also have direct benefits for ecosystems such as forests. Tropical forests are falling at the rate of approximately 1% per annum or 29 ha. per minute. Tissue culture of tree species is being used to rapidly multiply propagules of tree species that are difficult to produce by conventional methods or where the demand for seeds exceeds the supply. Some of the trees include Camphor, Mvule and Eucalyptus.

Recent research indicates that there have also been several field trials involving genetically modified (GM) trees. Since 1988, about 184 GM field trials of trees have been reported globally. Certain modifications are designed to use trees for bio-regeneration of polluted soil. GM trees are also being developed to reduce the current use of chemicals in the production of paper products. GM microorganisms can also be usefully introduced to control pollutants, tertiary oil recovery and frost damage. GMOs are already being used to convert nitrogen in the air into a form that plants can use to produce protein without the input of chemical fertilisers.

Tissue culture conservation procedures can also provide germplasm banks with tools for conservation of genetic resources. For instance in-vitro cultures of accessions, including those of endangered species, can be conserved and perpetuated for the future. Embryo transfer has been used to safeguard the conservation of some animals such as the Bongo, one of the largest of the forest antelopes, while cryopreservation of semen from elite bulls has been used in artificial insemination. In the field of pharmaceutical development genetically modified vaccines are now being used for control of livestock diseases.

Biological treatment of waste by microorganisms has been applied in the bioconversion and biodegradation of agricultural and industrial wastes. Bioconversion is based on the use of microorganisms or related enzymes geared towards converting the desired substances. Biotechnology has been used in the production of alternative energy sources through the conversion of agricultural and municipal waste leading into biogas, a source of energy based on anaerobic fermentation.

Necessity of Biosafety Measures

The need for adequate biosafety measures to protect the environment stems from the fact that the genetic modification technology is a relatively new technology, which raises — as with any new technology — questions about potential risks to the environment and human health. The limited experience with the technology also calls for efforts to apply the technology in the most judicious way so as to reduce the risk of any potential adverse effects on the environment or human health. In developing transparent systems, every country needs to create a system that takes into account the interest of all stakeholders.

A transparent system will involve the technology developers and the technology users and will have the ingredients of sound technologies backed by scientific information. A transparent system will take into account social and economic concerns and is based on priorities decided upon by the countries. This essentially calls for the development of national biosafety frameworks in consultation with all stakeholders and for the implementation of such frameworks in which workable and transparent systems are set up to handle requests for permits, monitoring and inspections, public information and participation.

The Biosafety Protocol is a next step in a process that started with UNCED towards the promotion of safe development and use of biotechnology. This protocol needs to be supported by national mechanisms that will have the necessary human and institutional capacity for its implementation. By setting transparent mechanisms for decision making, countries will be creating an enabling environment for the development and use of biotechnology. The UNEP-GEF project on Development of National Biosafety Frameworks and the projects on Implementation of National Biosafety frameworks endeavour to create the necessary capacity for developing countries so as to enable those participating countries to implement the Protocol and to fulfil their obligations under the Protocol.

The Biosafety Clearing-House

Indeed, everybody agreed that safe and transparent transboundary movement of living modified organisms (LMOs) would require an efficient global system to ensure the exchange of pertinent biosafety related data between countries developing and producing LMOs and countries importing and using them. Following this common view, the Biosafety Clearing-House (BCH) was established in Article 20 as one of the key tools to assist countries to implement the Cartagena Protocol. There was a common understanding that the BCH should take advantage of the most recent information communication technologies and should therefore be mainly Internet-based.

The development of the BCH then became one of the top priorities of the Intergovernmental Committee for the Cartagena Protocol (ICCP), the objective being to ensure its full operability at the date of entry into force of the Protocol. The Pilot Phase of the BCH was launched following the first meeting of the ICCP in Montpellier in December 2000, to build experience and provide feedback for the development of a functional and accessible Internet-based BCH as well as to identify and address the capacity needs of countries. Two years and two ICCP meetings later, version 2 of the Pilot Phase was released on February 1, 2003, and should serve as the basis for the transition to the fully operable BCH.

The structure of the BCH as it stands today is defined by its content, namely the different elements of biosafety-related information, as well as by its functions, which include mechanisms for registering and retrieving information. This is illustrated by the "Central Portal" and the "Management Center". The "Central Portal" offers a unique entry point to access the different elements of information, whereas the "Management Center" allows Parties to register and manage relevant national data. At the technical and technological level, the Pilot Phase of the BCH took advantage of the experience gained during the last ten years by the Secretariat as well as the Parties to the Convention on Biological diversity (CBD) in the development of the Clearing-House Mechanism (CHM) of the CBD.

In fact the Protocol explicitly establishes the BCH as part of the CHM. This integrated approach should ensure full compatibility between the two systems as well as efficient use of human and technological resources of the Secretariat. However, despite this close technical integration, the nature of information provided to the BCH is intrinsically different from that found on the CHM. In particular, information from the BCH will be used in legally-binding regulatory procedures and decisions, which has significant implication in terms of data quality control as well as compliance or liability.

The types of information accessible through the BCH can be divided into two groups: those mandated by the provisions of the Protocol and those that have been recommended by the governments through the ICCP to assist countries in implementing the Protocol. This second group includes the roster of experts on biosafety, the database on capacity-building needs and opportunities, a bibliographic search mechanism, Web links to relevant biosafety resources or even discussion forums. The information specifically mandated by the Protocol could also be classified in two categories that could be described as "importer-driven data" and "exporter-driven data".

The "importer-driven data" (e.g. national focal points, competent authorities, relevant biosafety legislation, national decisions, risk assessment reports) are provided by national governments to facilitate implementation of the Protocol and decision making under the Advance Informed Agreement procedure (AIA). The exporter driven data (e.g. the final decisions regarding the domestic use of LMOs that may be subject to transboundary movement for direct use as food, feed and processing according to article 11.1) are essential for the implementation of the procedure under Article 11 of the Cartagena Protocol. The BCH is organised on a decentralised basis, which should be adaptable to the high degree of variation in national capacities, infrastructure and needs.

Governments can register information directly to the central database through the Management Centre or by sending data to the Secretariat using more traditional means. Another alternative is the development of national or regional components of the BCH whereby information located on a remote database could be accessed through the Central Portal by users. This approach allows governments to keep full control of and responsibility for their own data, and to facilitate updating of that information. Common formats for the different sets of information have already been developed to facilitate the establishment of a network of databases that are interoperable with the Central Portal.

One of the major issues faced by the Pilot Phase of the BCH was the relatively limited amount of data provided by governments, which made it difficult to test in real conditions the tools and common formats developed for the Pilot Phase. To circumvent this situation and also to benefit from existing databases containing information on approved LMOs, close partnership was established with other international organisations such as the ICGEB and the OECD. Those partnerships resulted in access through the BCH to databases such as the bibliographic database of the ICGEB or the product database of the OECD and enabled the Secretariat team to design and successfully test interoperability between independent databases.

Future Challenges

The main challenge for the efficient operability of the BCH will be capacity building to ensure the full participation of all Parties and governments. As a first priority, all Parties to the Protocol should be able to register information directly to the central database through the Management Centre. The next step would be to develop national or regional databases and to aim to make such databases interoperable with the Central Portal. This will require significant and more targeted capacity-building efforts tailored to the needs of each country or region. Such efforts should not only include human resources, but also infrastructure, since many developing countries do not have sufficiently stable and rapid Internet connections to run standard information technology applications efficiently.

At the regional level, priority should be given to the development of regional nodes to either host information or provide technical support. Strengthening of capacity building requires additional financial resources and better use of existing resources. UNEP has therefore submitted to GEF a proposal on Building Capacity for effective participation in the BCH as an add-on module to the current GEF Project on development of National Biosafety Framework. Additionally, a transparent and effective co-ordination mechanism should be established under the supervision of the Secretariat to ensure efficient synergies with existing initiatives.

The BCH should be sufficiently flexible to adapt itself to the existing capacities, needs and requests from users in order to facilitate registration and retrieval of information. For example, some countries already have domestic information requirements related to the use of LMOs which are registered in national databases. Due to the regulatory system, the structure of the domestic database is often different from those of the BCH. This should be taken into account in order to avoid duplication of information as well as to promote interoperability. Along this line, it is also essential to develop tools such as thesaurai and controlled vocabularies to improve search facilities and to facilitate the use of different languages.

The BCH should also give access to information on all LMOs approved for use in the environment or as food, feed or for processing since these data are essential to achieve the objectives of the Protocol. This is an important issue since the legal requirements under the Protocol only apply to the LMOs approved after its entry into force. Therefore the ongoing collaboration with international organisations should continue to ensure access through the Central Portal to existing databases such as the OECD Product database. Similarly, countries that are not Parties to the Protocol should be encouraged to participate. Criteria should also be developed to allow participation of non governmental organisations, including industry.

The development of the Pilot Phase of the BCH in less than 4 years illustrates how the application of modern information management technologies can be used to build a powerful tool to assist the implementation of a multilateral environmental instrument. Such an impressive result was made possible in such a short time thanks to the full commitment of the Secretariat's staff and the support of the ICCP and its Bureau, as well as the members of the expert groups. The tool is there - the challenge will now be for all Parties, governments and relevant stakeholders to support the transition from the pilot phase to the fully operational and functional BCH by actively providing data and critically testing the BCH between 11 September and the first meeting of the Parties (MOP1) in February 2004 in Kuala Lumpur.

The success of this transition will allow MOP1 to take the necessary steps to reinforce the role of the BCH, not only as a tool to support the implementation of the Protocol, but more importantly as the first stone to build a global biotechnology information network to help achieve the objective of Agenda 21 towards sustainable development of modern biotechnology.

Biosafety Protocol in the National and International Levels

The issue of genetic engineering (GE) and biosafety has captured unprecedented public interest and concern around the world. Apart from the scientific discussion, disagreement and uncertainty surrounding genetic engineering, many other ethical and socio-economic concerns and realities are issues that have to be addressed in biosafety regulation at the international and national levels. At the national level, the implementation of the Biosafety Protocol is only one part of a wider and more comprehensive national system of biosafety regulation that countries put in place to meet the objectives of the Biosafety Protocol.

As the Biosafety Protocol is an international agreement of minimum standards for biosafety, the provisions regarding public participation are the minimum requirements that all Parties must fulfill. Parties can thus provide for wider and stronger participation, including the right to information, effective participation in decisionmaking and effective access to judicial and administrative proceedings. The last is particularly relevant in the formulation of a liability regime that will provide effective redress and remedy. There are valuable precedents in existing laws including the 'Aarhus' UN/ECE Convention on Access to Information, Public Participation in Decision-Making, and Access to Justice in Environmental Matters, and national environmental and social impact assessment procedures in many countries.

During the negotiations of the Biosafety Protocol, the active commitment and involvement by some sectors of the public, including non-governmental organisations (NGOs) and scientists helped to introduce valuable information, ideas and points of view that contributed positively towards the understanding of biosafety and the development of the Biosafety Protocol. Often, developing countries are at a disadvantage with poor information flows, and the multidisciplinary and holistic nature of biosafety requires the engagement of all sectors of society, including the wider scientific community, that may not normally be present in negotiating fora or regulatory bodies.

The process was also further complicated by the powerful trade interests of major producers of genetically modified organisms (GMOs) and associated products. It was effective public participation that greatly helped to maintain the integrity of the discussion on scientific, environmental, health and socio-economic issues. These experiences are also important at the national level. Public participation can be a valuable tool on many levels - in the development of the national biosafety law or system, in the decision-making on specific GMOs, and in the wider debate on genetic engineering and sustainable development in a particular country.

The latter is often not considered seriously at all but some countries, which introduced GMOs without consulting the public on these issues at the outset, have experienced a public backlash as a result. With limited resources and the uncertainties around even the claimed benefits of GMOs and associated products, developing countries need to be able to consider non-genetic engineering options to meet their developments priorities. These priorities necessarily require public discussion. The role of the public in developing countries in the areas of monitoring and enforcement should also be fully recognised given the limitations on resources.

The Biosafety Protocol in Article 23 recognises that for public participation to be meaningful, there must be access to information, and public awareness and education. A number of challenges present themselves. With polarised views and vastly divergent interests, a credible public awareness, education and participation process enables all views to be heard and debated on all issues, avoiding dishonest value judgments on the kinds of information and the way it is presented to the public. This also presents an opportunity to hear from all available expertise in the public from all fields, outside of the commonly used pool of expertise.

Experience has also shown that the public can well discuss scientific issues if these have been communicated in a way that people can understand, and that the public has an insightful wisdom in dealing with scientific uncertainty and science policy that must be heard if science is truly to serve the needs of society. Another key challenge in public participation is to reach out to the sectors of the public that are not organised, outside of the usual NGOs and civil society groups that are generally considered as representing the public views. These groups include farmers, especially the small farmers in developing countries, and consumers in general.

Integral to the implementation of effective public participation is access to information. Thus Article 21 on "confidential information" needs to be implemented in a manner that is consistent with Article 23 and the spirit and objectives of the Biosafety Protocol. During the Biosafety Protocol negotiations, the Africa Group (supported by a number of other developing countries) argued that a provision on protection of trade secrets and confidential business information was not necessary since these are already adequately protected under other related laws. The World Trade Organisation (WTO) Agreement on Trade-related Aspects of Intellectual Property Rights (TRIPS) already provides for such protection with prescribed criteria.

However, the Protocol provision remained at the insistence of some developed countries. Unfortunately it also evolved from "confidential business information" to just "confidential information" arguably going even beyond the TRIPS requirements. Thus Parties will have to be clear about defining the scope and conditions for claims for protection of confidential information in national biosafety laws. It is important to emphasize that the regulatory authorities have the right to all information necessary for decision-making and that Article 21 only refers to information that is confidential in relation to the public.

Experience reaffirms that it is only in an open and transparent system that all available information, views and assessment can be brought together to ensure sound and responsible collective decision-making. Since the possible consequences of hazards posed by genetic engineering to the environment and human health are so serious, information disclosure to the public - and especially scientists in the public domain - is crucial. There is already increasing public mistrust with the revelation of highly questionable corporate practices in recent years, and the growing number of cases where scientific dissent and evidence of hazards of genetic engineering are subject to intense pressure from industry.

Critical scientists have also been known to be personally vilified. Thus, to implement the spirit of Article 23, Parties have to design national laws and systems for sound technology assessment of genetic engineering with the involvement of the public at both the design and implementation stages.

Private Sector Involvement in the Implementation of the Cartagena Protocol

It viewed its role during the negotiations as providing concrete information on the realities of international trade in LMOs, existing international and national regulatory requirements and practices, and the potential implications of the various proposals under debate. The GIC offered this input in the context of obligations under the Convention on Biological Diversity (CBD) to promote technology transfer and cooperation and the distribution of the benefits of biotechnology, while at the same time providing an adequate level of safety, as part of the global quest for the conservation and sustainable use of biological diversity.

The role of the GIC, along with the International Grain Traders Coalition, the International Seed Federation, and other private sector organizations and associations concerned and involved with the Protocol, is even more critical at the implementation stage. Indeed, as the Protocol becomes reality for 50 countries on 11 September 2003, never has it been more important for the private sector to assist governments and the international community by providing detailed information on commercial practices, possibilities, and limitations based on its extensive experience with the regulation and trade of LMOs.

Workable, predictable, transparent, and effective regulatory structures are not only welcome, but also necessary for the private sector to operate. In the countries where biotechnology first was developed, the existence of flexible, science-based and effective regulatory mechanisms to ensure appropriate risk assessment and risk management of activities in this field were put in place early on. The Protocol, by offering a structure for ensuring science-based risk assessment of LMOs prior to intentional introduction into the environment, may offer the possibility for more countries to engage in appropriate and informed decision making and avail themselves of the benefits of biotechnology while providing for biosafety.

A lasting benefit of the Protocol for all concerned is the establishment of the Biosafety Clearing-House (BCH). A centralized mechanism that provides clear and accessible information about legislative requirements, the regulatory status of LMOs, risk assessment outcomes, and contact points for more information is critical for the regulated community, regulators and the public alike. Over time, it is anticipated that the BCH may lead toward greater harmonization and even more extensive information exchange. Of great interest will be the information ultimately posted by developing countries as their scientific institutes and research organizations identify and develop solutions to local challenges using modern biotechnology.

Key Challenges

As much potential as the BCH offers, key challenges in the implementation of the Protocol begin with the fact that the BCH is not yet fully operational for lack of information provided by countries Party to the Protocol. In some cases, information simply has not been posted; in others, there are clear misunderstandings about the nature of the information required. For example, many countries have provided detailed information about the various governmental bodies involved in biosafety regulation without providing a clear indication of to whom a notification should be sent.

The relationship between the legislation that can be found on the BCH and the Protocol is unclear in most cases because countries have not entered a declaration clarifying that the domestic legislation is intended to apply. Furthermore, given that it appears that legislation found in other databases has been imported into the BCH, it is unclear whether the legislation that appears is the correct and complete version. The private sector also is concerned that key implementation issues will not be resolved until February 2004 at the earliest.

Critical aspects of the Protocol with which the private sector ultimately will be expected to comply therefore will remain unknown for an extended period of time after entry into force of the Protocol. Effecting change in commercial practice requires the existence of clear expectations and requirements and a reasonable period of time to incorporate any new obligations. While some have suggested voluntary compliance with the recommendations of the Intergovernmental Committee on the Cartagena Protocol

(ICCP), it is evident that the Conference of the Parties serving as the Meeting of the Parties may or may not adopt these recommendations.

Further, it is unclear how issues that did not enjoy consensus during the ICCP process will be resolved. The resulting uncertainty creates a serious challenge for the private sector. In addition to uncertainty, the private sector is concerned about unrealistic expectations and interpretations of the Protocol at both the international and national levels. Interpretations and implementation decisions that exceed the requirements of the Protocol can result in serious negative effects on international trade, research and development in both developing and developed countries and, fundamentally, countries' access to food, feed, seeds, and other desired products.

The view that including "products thereof" in domestic implementing legislation for environmental safety somehow constitutes "stricter" legislation is one example. Such provisions clearly exceed the requirements of the Protocol; indeed, they were considered and rejected during the negotiations for good cause. Rather than offering more protection, inclusion of "products thereof" in domestic implementing legislation will inevitably overburden government regulators without environmental benefit. Forcing the development of new, technology-specific liability provisions in the absence of corresponding technology-specific risk is another aspect that would negatively impact technological development and even the environment.

Companies already are held liable for damage they cause, regardless of whether it involves the use of one technology or another. However, the establishment of biotech-specific liability regimes at the international or national levels would create disincentives against using modern, sustainable technologies. Such disincentives would include making the use of these technologies more expensive, legally risky even when they are shown to be at least as safe as conventional technologies, and difficult to insure, without enhancing the preventive effect of regulatory systems.

During the ICCP process, the private sector has been identified as a key stakeholder in Protocol implementation and a critical player in building the capacities necessary for countries to implement and comply with the Protocol. Indeed, the private sector is routinely called upon to contribute and must do so if the Protocol is to be workable and effective and technology transfer and cooperation as envisaged in the CBD and in Agenda 21 is to succeed. It is surprising, therefore, that some appear committed to circumscribing the role of the private sector rather than encouraging its active engagement. This was seen during the ICCP process, for example, when some governments proposed a limited role for the private sector in capacity building.

It is routinely seen when the private sector is called upon to assist with identified capacitybuilding needs but its efforts to respond directly to these requests are viewed with suspicion and sometimes even rejected. The ICCP process has been marked by increasing recognition of the important role of the private sector, alongside other stakeholders, on implementation issues. Further enhancing private-sector involvement in the implementation of the Protocol will require not only calling upon the private sector to act but an open mind and willingness to consider the contributions it offers. Providing more timely and direct opportunities for the private sector to offer its views on the potential implications of various options for implementing key Protocol proposals also would encourage greater private sector participation in the process.

Countries with functioning regulatory systems in place routinely engage in discussions with the private sector and other stakeholders about the implications of regulatory options under consideration. Building into the process additional platforms for these kinds of practical exchanges of views at the international level could foster a more practical, science-based approach that would ensure better decision-making on implementation matters.

Challenges and Opportunities of the Biosafety Protocol

For the Protocol to function in practice, countries need to have systems in place that allow them to process notifications, to carry out risk assessments, etc. This requires that countries establish what we call a "national biosafety framework", which includes a policy, a regulatory regime, a system to handle notifications, systems for monitoring and inspections, and systems for public information and participation. The establishment of a national biosafety framework is not something that suddenly became necessary because of the Protocol. Since 1992, Article 8g of the Convention on Biological Diversity (CBD) has called for the establishment of such national mechanisms. Yet, the coming into force of the Biosafety Protocol presents a new challenge.

On the one hand, countries need to establish a well thoughtthrough biosafety framework that is workable, transparent, and consistent with the Protocol and other international agreements. This process usually follows two phases: a development phase and an implementation phase. This two-phase approach is also reflected in the GEF Initial Strategy on Biosafety, which was aimed at assisting countries to be prepared for the coming into force of the Biosafety Protocol. It is important to recognise that although we can make use of over 20 years of experience with national biosafety frameworks, these two phases still will take several years.

On the other hand, and this is the challenge, countries need to be able to start "working with the Protocol" after they have become Parties. This challenge is similar to the task of someone who is planning to build a house, and must at the same time cook meals in a kitchen that mainly exists on the drawing board. Obviously, capacity-building projects face a similar challenge. The work of UNEP with regard to the GEF Initial Strategy on Biosafety consists of - among others:

— the UNEP-GEF Project on Development of National Biosafety Framework and

— the UNEP-GEF Projects on Implementation of National Frameworks.

Biosafety Capacity-building Initiatives

Having come across many biosafety capacity building projects over the last ten years, the following aspects such as: country ownership, duration, peer review, regular evaluation, forward thinking, and last but certainly not least -coordination, are important for the success of biosafety capacity building projects.

Country ownership

Projects need to be country driven, i.e. requested and executed by the countries. Organisations assisting in capacity building need to be available for assistance and guidance, but the real "drive" needs to come from the countries requesting the support. Projects that are 'forced upon' countries, often lead to nothing, because there may be no drive to get things done, and - worse — there may be no commitment for continuation.

Duration

Given the legal and scientific complexity of this field, and given the high 'turn over' of staff in the countries, long projects of several years are often preferable to a series of short projects.

Peer review

It is no help to countries if the support they are being given consists of an exotic approach that is not widely viewed as being valid and workable. It is therefore important to regularly present the approach taken to a review by peer experts. This is particularly the case for the scientific components of the implementation phase.

Regular evaluation

Biotechnology and biosafety evolve rapidly. It is therefore important to regularly re-evaluate the progress of a project, and adjust where necessary. While it is important to start a project with a clear plan, that plan should not be a straightjacket.

Forward thinking

The success of a capacity building project depends to a large extent on the continuation afterwards. This requires that part of the work should be aimed at 'follow up' when the projects are over. It is generally recognised that (sub) regional collaboration is one of the best mechanisms to foster continuity and maximise resources.

Coordination

Over the past few years, biosafety capacitybuilding initiatives have emerged in great numbers from a variety of sources. In many cases, the funding sources have been unaware of similar initiatives and there has been immense duplication of work that has wasted resources and sometimes was counter- productive.

That situation is slowly improving, but is still far from satisfactory. One of the most promising opportunities to avoid duplication is the Coordination Mechanism that has been called for by the ICCP and on which the CBD

Secretariat has taken some very useful initial actions. Key to any coordination of effort is information, and all involved in biosafety capacity-building are called upon to enter and update their information in the capacity building database of the BCH (http://bch.biodiv.org/ Pilot/Home.aspx). For the same reason, one of the first steps in the planning stage of any biosafety capacity-building project should be to consult the BCH to check whether similar initiatives are planned or being carried out already.

Identification of New and Innovative Funding Options and Delivery Strategies

Funding options

Presently, there are many planned and ongoing biosafety capacitybuilding initiatives funded by international and national organisations, Governments, NGOs and the private sector. The amount of money involved over the next five years is somewhere over US dollars 100 million.

Delivery strategies

If we don't act soon, one of the major challenges in biosafety capacity building will be the lack of experts who can act as trainers. At the moment, only a handful of people have substantial practical experience in developing and implementing national biosafety frameworks and are available and capable of giving training. Already now, with the growing number of projects, it is becoming increasingly difficult to get experienced people to be trainers in a workshop, and this problem will get worse over the next couple of years.

There is an urgent need to start searching for experts and then training those experts so that future demand will not far outstrip current supply. In addition, a clear strategy needs to be devised to address this problem in a more strategic and structured way, something that could be picked up by the liaison group that is part of the coordination mechanism mentioned above. When thinking about sub-regional collaboration, we should also be thinking about the longer term 'delivery strategy' of capacity building.

Future Implementation of the Biosafety Protocol

The Protocol is premised upon a system of functioning national biosafety frameworks and upon information exchange. Action at the international level is needed to ensure that all Parties are in a position to exercise their rights and meet their obligations under the Protocol, as well as to enhance the mutual understanding by Parties of those rights and obligations. The elaboration and clarification of the principles and rules enshrined in the Protocol by the Conference of Parties serving as the meeting of Parties to the Protocol (COP-MOP), the establishment of procedures and mechanisms to promote implementation, and cooperation between the Protocol and other relevant international agreements and organisations will all be central in achieving the Protocol's objective.

The Protocol specifies certain future work to be undertaken by the COP-MOP, and establishes timeframes for action. While some of these issues are of an administrative and budgetary nature, many others address implementation issues that were either not necessary or not possible, for reasons of a lack of consensus among governments or a lack of time, to articulate them fully in the Protocol itself. In addition to those items on which the COP-MOP is explicitly required to act, the mandate of the COP-MOP, in Article 29(4) of the Protocol, also incorporates "other issues necessary for the effective implementation of the Protocol". The issues on which the COP-MOP is required to take action at its first meeting include a core cluster of those related to capacitybuilding.

While a number of important capacity-building initiatives related to the Protocol are already underway, the early decisions to be taken by the COP-MOP on the basis of ICCP's deliberations will be important in laying the groundwork for the effective implementation of the Protocol. These include decisions on capacity-building, on facilitating decision-making under Article 10(7), on the role of the roster of experts, and on the recommendations to the CBD COP regarding guidance to the GEF. Closely related to this cluster of issues will be the decision on the functioning of the Biosafety Clearing-House (BCH), reliable access to which is central to all aspects of implementation of the Protocol, particularly the provisions regarding living modified organisms for direct use as food or feed or for processing (LMO-FFPs).

The first meeting of the COP/MOP will also have to establish certain processes and mechanisms that may elaborate or add to the body of substantive obligations of the Parties. In particular, processes will need to be considered to advance consideration of detailed requirements for documentation accompanying transboundary movements of LMO-FFPs under Article 18(2)(a), and to consider the appropriate elaboration of rules and procedures with regard to liability and redress. While both remain extremely contentious issues, it is important to the early credibility of the Protocol that processes are set in place that seek to complete work on these issues within the timeframes envisaged in the Protocol.

COP-MOP.1 must also consider and approve compliance procedures for the Protocol. For the foreseeable future, many compliance issues seem certain to remain inextricably linked to questions of capacity, for example to respond to and take decisions on notifications of proposed transboundary movements of LMOs. While some assistance will be available to Parties through the roster of experts and other mechanisms established under Article 10(7), such mechanisms cannot substitute for adequate legal frameworks and assessment, monitoring and decisionmaking capacities at the national level. The design and application of the compliance mechanism will therefore need to be sensitive and responsive to the capacity constraints faced by many Parties.

At the same time, they must be capable of addressing compliance issues and should therefore incorporate a variety of trigger mechanisms and response measures. In addition to this challenging agenda, there are numerous other issues which seem to merit additional consideration by the COP-MOP. Initial suggestions for questions that might be addressed under Article 29(4) have included: the categorisation of LMOs; risk assessment and risk management; establishment of harmonised rules for unique identification systems; and transboundary movements of LMOs involving non-Parties (ICCP recommendations 2/6 and 3/8). A number of other important implementation issues could be taken up here.

One might envisage the COP-MOP considering for example: how to ensure that or evaluate whether bilateral, regional or multilateral regional arrangement or agreements are 'consistent with the objective of the Protocol' and 'do not result in a lower level of protection'; what types of LMOs fall into the category of LMOs that pharmaceuticals for humans and are addressed by other relevant international agreements or organisations' for the purposes of Article 5; or further work to clarify relevant socio-economic considerations within Article 26. In addition to work that will take place in the COP-MOP, cooperation and coordination with other relevant international organisations will also play a key role in promoting effective implementation of the Protocol.

A flurry of international activity on biotechnology and biosafety in recent years has given rise to a complex body of relevant international rules and standards, some of which are still under development. International organisations and instruments relevant to the Protocol include those addressing agriculture, food safety and health, such as the Food and Agricultural Organization of the United Nations (FAO) and the International Plant Protection Convention (IPCC), the Codex Alimentarius and the World Health Organization (WHO). Efforts at coordination with these and other relevant bodies need to address both policy coordination — ensuring that states' rights and obligations regarding the regulation of LMOs are "mutually supportive" — and operational coordination.

At the policy level, the relationship that has attracted the most attention has been that between the Protocol and relevant agreements established under the World Trade Organization (WTO). Enhancing policy coordination and mutual supportiveness between these two bodies of international law requires ongoing two-way communication. In this regard it is important that the CBD is granted observer status in relevant WTO committees, and, of course, that representatives of the WTO and of other relevant intergovernmental organisations, continue to attend meetings held under the auspices of the Protocol. At the operational level, the CBD has by now extensive experience in a variety of mechanisms designed to facilitate cooperation between international organisations, including the establishment of memoranda of cooperation, and the development of targeted joint work programmes.

Such mechanisms can also be useful in promoting implementation of the Protocol and have already been taken up in the case of the IPPC in pursuance of ICCP recommendation 2/12. Operational cooperation will be of particular importance, for example, in the fields of information exchange and data management, particularly the functioning of the BCH, capacity-building, and implementation and enforcement mechanisms. In addition, as has already been the case in the drafting of the Protocol itself, there may be much that can be drawn from practice and procedures developed under other multilateral environmental agreements (MEAs), in relation, for example, to documentation requirements, reporting formats and monitoring. Of particular relevance here are other agreements which address the transboundary transfer of hazardous or potentially hazardous materials, such as the Basel and Rotterdam Conventions.

3

Biotechnology Development in India

After India became an independent nation in 1947, Prime Minister Jawaharlal Nehru's socialist vision to improve the quality of life for the masses identified a central role for science and technology. Given its limited resources for R&D in the past, the government encouraged process, rather than novel product innovations. This resulted in a strong pharmaceutical industry with capacities in bulk and generic manufacturing, low-cost process innovations and some novel drugs. India now has the twelth most successful biotechnology sector in the world as measured by number of companies. If the current rapid pace of progress of its health biotechnology industry is any indicator, India has a promising future at the national and international level.

Agricultural Biotechnology Research in India

Agriculture is the main occupation in India. The cultivation of land not only sustains their livelihood but also provides a social milieu for their day-to-day living. Around 35 years ago, agricultural production in India got a major boost with the introduction of dwarf varieties of wheat and rice. The introduction of these varieties led to a dramatic increase in the yields of the two crops. Some productivity enhancement came through the use of hybrids in corn, sorghum and millet although the area under cultivation of these crops has steadily decreased in the last decade.

In the last 10 years the yields of rice and wheat have also plateaued out. The productive agricultural areas in the North, due to continuous rice-wheat cultivation are encountering serious problems of sub-soil water depletion, deficiency of micronutrients in the soil and increase in the use of pesticides, fungicides and herbicides to control pests, pathogens and persistent weeds. Agricultural production is becoming more and more dependent on agrochemicals, thereby increasing input costs and causing significant damage to the environment and human health. Both farmers and consumers are at the receiving end -farmers by exposure to agrochemicals and consumers due to residues of agrochemicals in the consumed food.

While there is self-sufficiency in cereal grains at present, the yields and productivity of dryland crops, mostly grain legumes and oilseeds remain low and no major breakthroughs in productivity enhancement and yield stabilization have been achieved. Currently, India is importing both grain legumes and edible oils to meet internal demands. India's population is expected to reach approximately 1.5 billion by 2050. It is estimated that around 300 million (roughly 30%) of India's population is suffering from malnutrition. Due to socio-economic reasons, women and children are more at the receiving end in terms of micronutrient deficiency and low calorie intake.

Nutritional security for everyone would require more extensive availability of grain legumes, edible oils, fruits and vegetables, milk and poultry products. The challenges of malnutrition, enhanced productivity and crop diversification can be met by better resource management and by breeding more productive, more nutritious and at the same time less resource input demanding crops. Crop biotechnology, which broadly includes areas of development of transgenic crops, structural and functional genomics and marker-assisted breeding could provide us with the vital breakthroughs to achieve improvements in both quality and quantity in a sustainable manner.

With the advent of techniques of genetic engineering in the early seventies, the natural barrier to gene exchange has been removed. Sequences from varied sources like bacteria, viruses and eukaryotic systems can be transferred to plants to develop transgenic crop varieties. Achievements, to date, in plant biotechnology have surpassed all previous expectations and with the development of high throughput instruments, the future is even more promising.

Recognizing the importance and potential of biotechnology, the Government of India set up the National Biotechnology Board in 1982 which became a full-fledged department (called as Department of Biotechnology, DBT) in 1986 under the Ministry of Science and Technology. To provide research and human resource development in crop biotechnology, DBT supported the establishment of Centres for Plant Molecular Biology (CPMB) in 1990. A total of six such centres were set up initially at various Universities/ Institutions namely, Jawaharlal Nehru Univer-sity (New Delhi), Madurai Kamaraj University (Madurai), Tamil Nadu Agricultural University (Coimbatore), Osmania University (Hyderabad), National Botanical Research Institute (Lucknow) and Bose Institute (Kolkata). Seventh centre was established at the University of Delhi South Campus in 1997.

A total financial outlay Rs 45.98 crores was made available for the development and research activities of CPMBs. Most of the research on crop biotechnology currently underway in India is being supported by the Government and the DBT has been the main agency amongst government departments to support research endeavours in this area. The DBT has supported over the last 12 years large number of research projects which deal with the development of in vitro regeneration and genetic transformation protocols of important crop species grown India, the development of transgenics with genes agronomic importance and several projects dealing with basic research in plant molecular biology.

To bring investigators with expertise in different facets of biology together to address defined research goals, number of multi-institutional projects have been launched which include development of transgenics for resistance to geminiviruses in cotton, mungbean and tomato, resistance to rice tungro disease, resistance to bollworms cotton, development of nutritionally improved potato with a balanced amino acid composition using a heterologous protein from Amaranthus and development of molecular methods for heterosis breeding. Another set of projects include tagging of quality traits in rice and mustard, resistance to karnal bunt and rust in wheat and extensive research network on rice biotechnology, which includes work on genomics, mapping, improvement nutritional quality and resistance to biotic and abiotic stresses.

Another multi-institutional project on tagging quality traits in wheat by molecular markers has given some interesting leads. A number of markers linked grain protein content, seed size, bread-making quality, etc. have been identified. To further strengthen research in the area of crop biotechnology, a new institute, National Centre for Plant Genome Research (NCPGR) has been established in New Delhi with a mandate to strengthen plant biotechnology research in India.

Rice production in India holds the key to the food security of the country. After detailed consultations with experts at the National and International level, DBT and the Indian Council of Agricultural Research (ICAR) jointly took the initiative to participate in the international rice genome-sequencing programme (IRGSP) in June, 2000. Two laboratories in India, one at the University of Delhi South

Campus and the other at the Indian Agricultural Research Institute (IARI) in New Delhi are jointly sequencing part of chromosome II (region between 57.3 cM and 110.1 cM of the genetic map). This region carries genes for a number of biotic and abiotic stresses and genes encoding some agronomically important traits.

Around 15.58 Mb of DNA has already been sequenced till December 2002. A high quality draft sequence of rice genome by IRGSP has been declared completed by the Japanese Prime Minister Junichiro Koizumi on 18 December 2002 in Tokyo. In order to take advantage of sequence information in rice, a workshop was organized in May 2002 to identify areas which should be funded for functional genomics of rice. It was decided to develop research progammes in the areas of:

(i) salinity and drought stress,

(ii) biotic stresses,

(iii) mapping of loci contributing to high yield.

The project on genome sequencing will also allow rapid progress in the characterization of allelic diversity available in Oryza sativa and related species and genera for resistance to diseases and pests of rice.

Future Developments

In the 10th Five Year Plan (2002-2007) the following areas will be given high priority:

— Rice genomics with particular emphasis on the study of genetic variability for achieving both yield increase and stabilization in the crop.

— Molecular maps in rice, wheat and grain legume crops. To date, a molecular map with reasonable density (around 1000 AFLP markers) has been developed only for mustard (Brassica juncea). In wheat, tagging of some interesting traits has been accomplished. The work on cereal crops will continue and mapping work will be initiated in grain legume crops. Molecular maps will be used for mapping of QTLs for abiotic stresses, plant architecture, and yield and mapping of loci conferring resistance to pests and pathogens.

— Transgenics developed in various crops through DBT funding will be tested under appropriate containment conditions and if found useful will be subjected to nutritional and toxicological tests to determine their suitability for cultivation in the farmer's field.

— Research on development of pollination control methods for the production of hybrid seed will be further strengthened. Fixation of heterosis through apomixis will be another area of research.

— Particular emphasis will be given on an improvement in the nutritional status of some of our major crops. To begin with, the three target crops will be rice, wheat and potato. High-yielding rice varieties grown in India will be either transformed with constructs that would code for high b carotene content or by back-crossing these varieties to 'Golden rice' transgenics already developed in Switzerland. Improving the iron content in rice grains will be another thrust area. In potato, nutritionally enhanced transgenic Indian varieties of potato carrying the ama1 gene are already under advanced field trials. If found suitable, these will be released for cultivation.

— Transgenic approaches to senescence retardation in fruits will be a major thrust area. It is important to reduce post-harvest losses in fruits and vegetables as under tropical and sub-tropical conditions senescence is much more rapid and energy costs of cold-storage are very high.

Biosafety Considerations

Safety of transgenic material is a major concern. A major

contribution of DBT has been to set up, in collaboration with other government departments, a framework for the evaluation and eventual clearance of transgenic crops for field cultivation. The release of transgenic crops is governed by the Indian Environment Protection Act (EPA) - 1986 which came into force from May 1986. The act provides a framework for the protection and improvement of environment. Later, the rules and regulations for the manufacture, use, import, export and storage of hazardous microorganisms, genetically engineered organisms or cells were notified under EPA 1986 on 5 December 1989. A mechanism based on interaction between committees and different departments of Government of India has been set up.

Such materials will have to meet with the approval of the following Committees: Institutional Biosafety Committee (IBSC), Review Committee on Genetic Manipulation (RCGM) and Genetic Engineering Approval Committee (GEAC). Under the EPA 1986, the GEAC examines from the viewpoint of environmental safety and issues clearance or the release of genetically engineered organisms/ transgenic crops and products into the environment. The GEAC also grants permits to conduct experimental and large-scale field trials which are beyond the limit of 20 acres.

In case of transgenic crops, applicants are required to seek clearance from the Ministry of Agriculture, which will accord the final clearance of transgenic crop varieties for agricultural activities. Bt cotton was released for commercial cultivation in March 2002, following the biosafety procedures laid down by the Government of India. Three cotton hybrids have been granted permission for field sowing and are currently in the fields in six states namely, Maharashtra, Gujarat, Madhya Pradesh, Andhra Pradesh, Karnataka and Tamil Nadu. Transgenic mustard hybrids are also in advanced field trials and if found suitable, may be released for general cultivation.

Health Biotechnology Sector

The Shanvac-B vaccine uses novel expression technology and is an important success story for India's health biotechnology sector. Shantha Biotechnics (Hyderabad, India), a biotechnology start-up, began research for an affordable indigenous vaccine in 1993 in a government-sponsored collaborative program with Osmania University (Hyderabad, India) and with financial support from Oman under the patronage of His Excellency, Yusuf Bin Alawi Abdullah, Foreign Affairs Minister of Oman.

Subsequently, scientists from the Centre for Cellular and Molecular Biology a national laboratory of India's Council for Scientific and Industrial Research (CSIR, New Delhi) and various hospitals joined the project to successfully develop Shanvac-B within four years. Compared with previous imports that cost approximately $16 per dose, Shanvac-B sells for about 50 cents in India. Although India has yet to introduce a novel health biotechnology product, it has a strong science base from which to launch innovations.

Its indigenous 'me-too'products take advantage of local strengths, focusing on low-cost processes and relatively simple vaccine, therapeutic and diagnostic products. Moreover, these products address Indian health needs. Even so, Indian entrepreneurs are also looking to global markets. As one respondent from the private sector explained, "...If you are thinking global, you are automatically dreaming big and you are automatically... getting a bigger share out of it." Echoing this sentiment, India's private sector is leveraging its capacities in affordable biopharmaceuticals to establish an international presence.

Firms like Bharat Biotech International (Hyderabad, India), for example, have obtained research and product development grants from international sources, including the Bill and Melinda Gates Foundation (Seattle,WA, USA) and the European Malaria Vaccine Initiative (The Hague). In 2002, Shantha secured World Health Organization (WHO; Geneva) certification, and a contract with the United Nations Children's Fund (UNICEF; New York) for 8.5 million doses. Two other companies, Panacea Biotec (New Delhi) and Serum Institute of India (New Delhi), are also suppliers to UNICEF. This indicates an interest in taking products developed for local needs, as seen in the hepatitis B case, and tapping into the large global markets.

A comparison of India's health biotechnology publications in international peer-reviewed journals and patents granted by the United States Patent and Trademark Office (USPTO; Washington, DC, USA) between 1991 and 2002 provides another view of the country's innovation level in terms of scientific output and commercial potential in the field. Data derived from Science-Metrix show India's scientific publications have been increasing quite steadily in health biotechnology. Its patent activity based on inventors' addresses in USPTO-granted patents in health biotechnology also shows an increasing trend. The year 2002 showed the highest number of publications and patents granted for this period.

Both national and state governments have made substantial investments in biomedical research as a means of tackling India's unmet medical needs and improving the overall health of the population. This has fostered the growth of well-developed academic and research institutions, which in turn support a private sector that is active both domestically and internationally. Private investments for biotechnology ventures are also beginning to emerge. The Dynam Ventureast Group and Andhra Pradesh Industrial Development Corporation (APIDC) recently established a joint venture (APIDC-VCL) to create India's first national VC fund for biotechnology. The Indian population is almost universally positively disposed toward healthcare biotechnologies.

Governments: Although health is largely under the purview of state governments, the national government also oversees National Health Programs focusing on communicable or poverty diseases, including malaria, tuberculosis, AIDS, leprosy, filaria and iodine deficiency disorders. The respondents identified the national government, and to some extent state governments, as the main source of funding in health biotechnology. Projects by India's extensive public R&D network are prioritized by public health needs. National government allocations for biotechnology in 2002-2003 were 2,210 million rupees (~$47.9 million), rising to 2,730 million rupees (~$59 million) in 2003-2004.

Many state governments, such as Karnataka and Andhra Pradesh also provide funding for R&D, which includes supporting health biotechnology. India's sixth Five Year Plan (1980-1985) identified biotechnology as a means to address the developmental needs of its agricultural and health sectors. This marked a long-term commitment by the government to use science and technology as a tool to tackle India's underdevelopment and to improve public health. The Department of Biotechnology (DBT;New Delhi) was created in 1986 to promote agricultural and healthcare biotechnology sectors through administrative measures and policy formulation.

The DBT provides support in the form of human resource development, biotechnology facilities and bioinformatics programs. In terms of research, the DBT's priority areas in health biotechnology include developing new generation vaccines for cholera, rabies, Japanese encephalitis, tuberculosis, malaria and HIV infections under the National Jai Vigyan science and technology mission. Other government agencies are also playing roles in policy making and in providing coordination and support for diverse R&D projects to develop vaccines and diagnostics for numerous diseases.

For example, the Indian Council of Medical Research (ICMR; New Delhi) is responsible for formulating, coordinating and promoting biomedical research to deal with national health priorities. Its network of 21 research institutes and centers, as well as six regional medical research centers, carries out studies in specific areas, such as cholera, AIDS, tuberculosis and leprosy, as well as region-specific health problems like the health of geographicallydistributed Indian tribes (most being relatively disadvantaged, and some still adhering to primitive practices).

The ICMR also funds various extramural projects that meet its mandate. Under the Department for Scientific and Industrial Research (New Delhi), the CSIR is responsible for a wide range of scientific and industrial research projects, from aerospace to mineral mining, and from designing animals and plants as bioreactors to coal preparation. Like the ICMR, it has a strong support structure with 40 laboratories and institutes, and 80 centers for technology transfer and other R&D-based activities. Although the CSIR oversees several health biotechnology projects, including the application of molecular techniques to drug target and predictive marker discovery, the CSIR is not solely focused on health-related research nor is it dedicated to the field of biotechnology.

Research institutes and universities: Research institutes, research laboratories and universities are critical to India's health biotechnology sector. They not only provide the scientists and technicians for the sector's workforce, but also contribute research discoveries of relevance to biotechnology firms and others in the field. For example, according to the USPTO health biotechnology patent data for India, in 2002 the CSIR was the predominant assignee. This is representative of the successes in its research efforts. It provides licensing services-with a focus on potential industrial partners-from its intellectual property. It does this through its intellectual property management division, Patestate.

Traditionally, the role of universities was in education and training, whereas laboratories, both public and private, focused on research. Today, the lines are blurring, as universities incorporate research activities and national laboratories provide training for students. A fine example of the blurring of these functions is seen in the Indian Institute of Science. The IISc was established in Bangalore in 1909 and became a university in 1956. The institute incorporates both teaching and laboratory research. It has a division of biological sciences, consisting of various departments, a center and a unit, all of which are involved in diverse research projects, many relevant to health.

For example, the department of biochemistry is working on immunology, reproductive biology and plant development as part of the study of such diseases as malaria, rabies and tuberculosis, as well as carrying out applied research on drug targets and vaccines. Of the 15 most active Indian universities publishing in the health biotechnology field in internationally peer-reviewed journals for the period 1991-2002, the IISc ranked first with 20.5% of the total publications from these 15 universities coming from the IISc.

With a combination of innovative research and education, the IISc provides a good model for universities in India. Although universities contribute most in terms of Indian health biotechnology publications in internationally peer-reviewed journals-accounting for 50.6% of all Indian health biotechnology publications between 1991-2002-public research institutes follow closely behind with 43.3% of total health biotechnology publications from the country. The 1990s saw a flourishing of numerous institutes and laboratories dedicated to biotechnology-focused research under government departments like the DBT, ICMR and the CSIR.

Some of the most active in health biotechnology publications include the National Institute of Immunology (New Delhi), CCMB and the Institute of Microbial Technology (Chandigarh, India). Of the 15 most active Indian public research institutes publishing in the health biotechnology field in internationally peer-reviewed journals for the period 1991-2002, the CCMB published 5.8% of the total publications. It covers basic and applied R&D, including biomedicine and diagnostics, gene regulation in prokaryotes and eukaryotes, host-parasite interactions, membrane biology, protein structure, bioinformatics and functional genomics. The United Nations Educational, Scientific and Cultural Organization recognizes it as a 'center of excellence' and the Third World Academy of Science (Trieste, Italy) deems it a 'center for excellence for research and training.'

Industry. Local firms, both public and private, are playing a leading role in promoting health biotechnology. According to a report published in 2003, over 328 companies and 241 institutions in India use some form of biotechnology in agricultural, medical or environmental applications. Although estimates of its size vary, the domestic health biotechnology market accounts for about 55% of the total biotechnology market in India. But many Indian firms also export to foreign markets or provide services to foreign customers. The combination of a well-established established generic/bulk pharmaceutical industry that is diversifying and becoming successful in biotechnology endeavors, together with the emergence of dedicated biotechnology firms in the 1990s, has provided India with competitive strength in biopharmaceuticals.

The Ernst & Young report identifies 96 enterprises exclusively as biotechnology companies, making the Indian sector the third largest in the Asian region behind Australia (228 enterprises) and China/Hong Kong (136 enterprises). The sector is a diverse mix of private domestic small- and medium-sized enterprises, such as Shantha Biotechnics and Bharat Biotech International, larger firms like Biocon (Bangalore, India), which successfully floated on the market in spring, as well as multinational enterprises such as Dr. Reddy's Laboratories (Hyderabad, India) and Ranbaxy Laboratories (New Delhi, India). There are also a handful of public enterprises including Haffkine Bio-Pharmaceutical (Mumbai, India) and Indian Immunicologicals (Hyderabad, India).

Several multinational pharmaceutical giants, such as Eli Lilly (Indianapolis, IN, USA), GlaxoSmith Kline (Brentford, UK), Novartis (Basel, Switzerland), Pfizer (New York) and AstraZeneca (Wilmington, DE, USA), also carry out manufacturing and clinical trials in India, but they play only a minimal role in developing the local health biotechnology sector. Elsewhere in the health biotechnology sector, a few companies also focus on bioinformatics, including Strand Genomics (Bangalore, India), Infosys (Bangalore, India), Tata Consultancy Services (Mumbai, India) and Ocimum Biosolutions (Hyderabad, India).

Besides biopharmaceuticals, India is also poised to become a leader in biotechnology services, such as contract research. It has the advantages of lower costs of operations and manufacturing than many nations of the Organization for Economic Cooperation and Development (OECD; Paris), a large and relatively inexpensive scientific and technical workforce that speaks English, expertise in software technologies and a well-developed public R&D infrastructure. Successful contract research firms include Biocon subsidiary Syngene (Bangalore, India) and SIRO Clinpharm (Mumbai, India).

Given a wide range of skills and a move to research collaborations, strategic alliances, joint ventures, exports and public-private partnerships, India's health biotechnology sector appears poised for exponential growth. In India, modern venture capital did not exist until the mid-1980s.

Private investors come primarily from banks, financial institutions and insurance companies, including the ICICI (Industrial Credit and Investment Corporation of India, Hyderabad, India), the Small Industries Development Bank of India, (SIDBI, Lucknow, India and New Delhi) and the IFCI Venture Capital Funds (New Delhi) However, in terms of biotechnology, there are concerns that venture capital is "mostly available to companies whose product and market are clearly identified and research leads are already available for commercialization".

However, recognition in the country is growing of the importance of VC, especially for the early stages in biotech ventures. In 2003, the first national VC fund for biotechnology in India (The Biotechnology Venture Fund) was initiated with a joint venture, APIDCVCL, between Dynam Ventureast Group and Andhra Pradesh Industrial Development Corporation (APIDC, Hyderabad, India). In 2004, it provided about R 80 ($17.7 million) crore in funding to biotech firms. Given the reluctance of existing venture capital funds to invest in early stage developments, APIDCVCL is targeting early stage tech businesses.

The general public: According to the majority of the respondents, a significant portion of the population does not understand the implications of health biotechnology products compared with those of agricultural biotechnology products. The former are accepted readily without much debate about their risks and benefits, whereas the latter are viewed with more skepticism, largely as a result of anti-biotechnology activist campaigns in India and abroad. Some respondents also spoke of a cultural receptivity for science in India. The proscience attitude instilled in post-independent India by leaders like Nehru has resulted in a general trust towards high technologies, such as health biotechnology.

Given this confidence in scientists, doctors and others involved in health biotechnology, there is a responsibility to see that this faith is not undermined by scandal or scares. As one respondent noted when commenting on the general population's ignorance of health biotechnology, "The responsibility is much higher on the sciences to do good rather than exploit them [the people]." India's large population with its ethnic variation is a valuable resource in its health biotechnology developments.

Challenges in Development

Despite the rapid growth of its health biotechnology sector, India faces challenges if it is to encourage further startup creation. Two clear challenges are to streamline its regulatory environment and to address the issues faced by industry with the change to India's intellectual property laws.

A daunting regulatory system: The health biotechnology industry faces a cumbersome regulatory environment affecting clinical trials for drug approval, and regulated drug costs due to uncertainties about government policies on regulation. When asked about the role of the regulatory system in health biotechnology, the overwhelming concern was over the process of regulatory approval. As one respondent from the private sector remarked, the regulatory system "is impeding" development.

Introducing biotechnology products into the domestic market requires approval from several statutory bodies. For example, the Drug Controller General of India, which falls under the Ministry of Health and Family Welfare (New Delhi), deals with biopharmaceutical products after an institutional biosafety committee (each institute requires an institutional biosafety committee to assess research proposals) and the Review Committee for Genetic Manipulation (New Delhi) clear them. However, these products must also receive final clearance from the Genetic Engineering Approval Committee under the Ministry of Environment and Forests (New Delhi) to deal with concerns about possible environmental harm. The way the current regulatory system operates is creating confusion, stagnation and frustration for those involved in biotechnology research and in product development. For example, approval of Shanta's streptokinase product had been delayed by almost 10 months. Biocon also had setbacks in gaining approval for its insulin.

A patent system in transition: The Indian Patent Act (IPA) of 1970 is perhaps one of the most widely noted, long-lasting and farreaching influences on health biotechnology innovation in India. With limited resources for R&D, the IPA was created to encourage process patenting rather than novel product development. This act allowed the pharmaceutical industry to reverse engineer products that were still under patent protection. This enabled local industry to build up capacity in cost-effective process innovations and to target generic drug manufacturing. The weak patent system has had a significant impact on innovative capacities because it has emphasized process innovation, rather than product creation.

Because India is to become compliant with the World Trade Organization's (Geneva, Switzerland) Agreement on Trade-Related Aspects of Intellectual Property Rights (TRIPS) by 2005, significant changes are likely in the recognition of product patents, longer patent terms, and stronger compliance and enforcement mechanisms. Two amendments have already been made to the IPA and a third one, which would make it TRIPS compliant, is to be reviewed by the Indian government. Pressure to comply with TRIPS is raising awareness of intellectual property (IP) issues in the private and public sectors. For example, the private sector is increasingly investing more resources into R&D for novel product development.

Two cases illustrate the Indian government's approach toward the solution of IP rights issues. In a first case, the CSIR led a successful appeal against the USPTO's patent for turmeric in wound healing. This precedent illustrated India's capability and resolve to protect its traditional knowledge. Limitations in the ability of patent regimes in developed countries to protect traditional knowledge have spurred an Indian initiative to create a digital traditional knowledge database to curtail further instances of foreign patenting of India's biodiversity. Second, in 2002, the ICMR adopted a new IP rights policy that promoted public-private partnerships to encourage the development of health technologies. This effort hinges on emphasizing research that will lead to patents.

India's IP law is thus in transition. Its increasingly stringent IP protection reflects the industry's transition from process innovation and generic manufacture to product innovation and a desire to become a global player. The biggest challenge is whether under this transitioning patent system, India's strong pharmaceutical base, which is diversifying into biotechnology, and its current biotechnology sector will be able to cope with the stronger patent system and focus more on R&D for truly innovative products as opposed to focusing on generics, 'me-too' products, and process innovations.

4

National Biotechnology Development Strategy

Section I
Introduction

Biotechnology, globally recognized as a rapidly emerging and far-reaching technology, is aptly described as the "technology of hope" for its promising of food, health and environmental sustainability. The recent and continuing advances in life sciences clearly unfold a scenario energized and driven by the new tools of biotechnology. There are a large number of therapeutic biotech drugs and vaccines that are currently being marketed, accounting for a US$40 billion market and benefiting over a hundred million people worldwide. Hundreds more are in clinical development. In addition to these there are a large number of agri-biotech and industrial biotech products that have enormously helped mankind.

The Indian Biotechnology sector is gaining global visibility and is being tracked for emerging investment opportunities. Human capital is perceived to be the key driver for global competitiveness. Added to this is a decreasing appetite for risk capital in developed countries, which has led to a decline in the biotechnology sector in these regions where survival lifelines are being provided by the lower cost research environs of the developing world such as India. For a country like India, biotechnology is a powerful enabling technology that can revolutionize agriculture, healthcare, industrial processing and environmental sustainability.

The Indian biotechnology sector has, over the last two decades, taken shape through a number of scattered and sporadic academic and industrial initiatives. The time is now ripe to integrate these efforts through a pragmatic National Biotechnology Development Strategy. It is imperative that the principal architects of this sector along with other key stakeholders play a concerted role in formulating such a strategy to ensure that we not only build on the existing platform but expand the base to create global leadership in biotechnology by unleashing the full potential of all that India has to offer.

Why Biotechnology is Important to India

Biotechnology can deliver the next wave of technological change that can be as radical and even more pervasive than that brought about by IT. Employment generation, intellectual wealth creation, expanding entrepreneurial opportunities, augmenting industrial growth are a few of the compelling factors that warrant a focused approach for this sector.

Vision & Mission:

Biotechnology as a business segment for India has the potential of generating revenues to the tune of US$ 5 Billion and creating one million jobs by 2010 through products and services. This can propel India into a significant position in the global biotech sweepstakes. Biopharmaceuticals alone have the potential to be a US$ 2 billion market opportunity largely driven by vaccines and biogenerics. Clinical development services can generate in excess of US$1.5 billion whilst bioservices or outsourced research services can garner a market of US$1 billion over this time scale. The balance US$500 million is attributable to agricultural and industrial biotechnology.

India has many assets in its strong pool of scientist and engineers, vast institutional network and cost effective manufacturing. There are over a hundred National Research Laboratories employing thousands of scientists. There are more than 300 college level educational and training institutes across the country offering degrees and diplomas in biotechnology, bio-informatics and the biological sciences, producing nearly 500,000 students on an annual basis. More than 100 medical colleges add ~17,000 medical practitioners per year. About 300,000 postgraduates and 1500 PhDs qualify in biosciences and engineering each year. These resources need to be effectively marshaled, championed and synergized to create a productive enterprise.

India is reorganized as a mega bio-diversity country and biotechnology offers opportunities to convert our biological resources into economic wealth and employment opportunities. Innovative products and services that draw on renewable resources bring greater efficiency into industrial processes, check environmental degradation and deliver a more bio-based economy. Indian agriculture faces the formidable challenge of having to produce more farm commodities for our growing human and livestock population from diminishing per capita arable land and water resources. Biotechnology has the potential to overcome this challenge to ensure the livelihood security of 110 million farming families in our country.

The advancement of biotech as a successful industry confronts many challenges related to research and development, creation of investment capital, technology transfer and technology absorption, patentability and intellectual property, affordability in pricing, regulatory issues and public confidence. Central to this are two key factors: affordability and accessibility to the products of biotechnology. Policies that foster a balance between sustaining innovation and facilitating technology diffusion need to be put in place. There are several social concerns that need to be addressed in order to propel the emergence of biotechnology innovation in our country such as conserving bioresources and ensuring safety of products and processes. Government and industry have to play a dual role to advance the benefits of modern biotechnology while at the same time educate and protect the interests of the public. Wide utilization of new technologies would require clear demonstration of the new added value to all stakeholders.

The National Science and Technology Policy of the Government and the Vision Statement on Biotechnology issued by the Department of Biotechnology have directed notable interventions in the public and private sectors to foster life sciences and biotechnology. There has been substantial progress in terms of support for R&D, human resource generation and infrastructure development over the past decade.

With the introduction of the product patent regime it is imperative to achieve higher levels of innovation in order to be globally competitive. The challenge now is to join the global biotech league. This will require larger investments and an effective functioning of the innovation pathway. Capturing new opportunities and the potential economic, environmental, health and social benefits will challenge government policy, public awareness, educational, scientific, technological, legal and institutional framework.

The issue of access to the products arising from biotechnology research in both medicine and agriculture is of paramount importance. Therefore, there should be adequate support for public good research designed to reach the unreached in terms of technology empowerment. Both "public good" and "for profit" research should become mutually reinforcing. Public institutions and industry both have an important role in the process. The National Biotechnology Development Strategy takes stock of what has been accomplished and provides a framework for the future within which strategies and specific actions to promote biotechnology can be taken. The policy framework is a result of wide consultation with stakeholders - scientists, educationists, regulators, representatives of society and others and reflects their consensus.

It focuses on cross-cutting issues such as human resource development academic and industry interface, infrastructure development, lab and manufacturing, promotion of industry and trade, biotechnology parks and incubators, regulatory mechanisms, public education and awareness building.

This policy also aims to chalk out the path of progress in sectors such as agriculture and food biotechnology, industrial biotechnology, therapeutic and medical biotechnology, regenerative and genomic medicine, diagnostic biotechnology, bio-engineering, nano-biotechnology, bio-informatics and IT enabled biotechnology, clinical biotechnology, manufacturing & bio-processing, research services, bio-resources, environment and intellectual property & patent law.

Several state governments have enunciated biotech policies spelling out a comprehensive blueprint for the sector. It is, therefore, prudent to have a National Biotech Development Strategy that charts an integrated 10-year road map with clear directions and destinations. This is the time for investment in frontier technologies such as biotechnology. It is envisaged that clearly thought-out strategies will provide direction and enable action by various stakeholders to achieve the full potential of this exciting field for the social and economic well being of the nation.

Section II
Key Policy Recommndations and Interventions

2.1 Human Resource Development: Academic and Industry Needs

Biotechnology is a knowledge-driven technology, which needs to be driven by a flow of new ideas and concepts in the development of new tools for research, new processes for manufacturing and innovative business models. Rapid responses are needed to meet the challenges as they unfold and there is a requirement for specialized personnel and centres of excellence for R&D. The policy goal for the next decade is to facilitate the availability of scientific and technical human resource in all disciplines relevant to the life science and biotechnology sector. In order to build a successful biotechnology sector, large talent pools are required in multiple scientific disciplines such as molecular and cell biology, chemistry, physics, engineering, bioinformatics, medicine, agriculture, microbiology, technology transfer & commercialization, bioentreprise & biofinancing and intellectual property rights management.

Product and process development are inter-disciplinary in nature and deficiencies in specific areas may weaken the whole sector. The key issue is the manner in which to create an effective interface across disciplines. Reliable estimates of human resource availability for the next 10 years are required. Expert consensus indicates that there is adequate enrollment currently at the post-graduate and under-graduate levels, however the quality is inconsistent. Areas such as intellectual property rights, regulatory issues and industrial training have received inadequate attention. There is a consensus that there is an urgent need to augment the number of Ph.D. programs in the Life Sciences and biotechnology. A strong pool of academic leaders is key to sustained innovation.

Strategic actions:

(i) National Task Force on education & training

— A National Task Force will be created to formulate model undergraduate and postgraduate curricula in Life Sciences keeping in view, future needs. The said curricula must address the underlying need for multidisciplinary and inter-disciplinary learning and the appropriate stage for biotechnology training.

(ii) Need assessment

— There would be need assessment in 2005 for the next five years and close monitoring during the period for interim changes.

— A 10-year perspective plan for human resource will be prepared every five years.

(iii) Curriculum development

— Course curricula will be reviewed and improved in consultation with industry and research establishments and standard e-learning modules will be developed for specific skill areas such as IPR, regulations, and bioentreprise.

— Hands on exposure to M.Sc. biotechnology students will be enhanced through an extended industry internship as well as through short-term placements at CSIR and other appropriate National Institutes.

— Dual degree programs in biotechnology that include regulatory matters, IPR and bio-enterprise management will be encouraged and supported by the Department of Biotechnology.

— Emphasis will be given on training of high quality technicians and technologists in skills required by the industry by establishing Regional training centres at diploma, graduate and postgraduate levels.

(iv) Quality improvement

- An accreditation mechanism will be put in place for ensuring minimum standard of education and training at the post graduate and undergraduate levels. Base requirements for teaching and laboratory infrastructure will be specified and enforced.
- Teachers training programs will be taken up by creating regional teachers training centres

(v) Strengthening of teaching and R&D in life sciences and biotechnology in the university system

- Strengthening R&D in Life sciences and biotechnology in the university system will be accorded high priority. This is considered important for improving the quality of education and providing exposure to new technologies for students at various levels. Specific mechanisms to achieve the goal will include
- Creation of inter-disciplinary centres of excellence with world class infrastructure in key areas
- Program support to encourage inter-departmental networking
- Visiting professorship and creation of industry sponsored chairs in partnership with the Department of Biotechnology.

(vi) Attracting talent to life science and biotechnology

- Bright students will be attracted to take up careers in biology and biotechnology through special scholarships. Summer assignments at Academic and Industry research laboratories will be introduced at the school level to create interest in the fields of biotechnology and biology
- Women scientists will be encouraged to take up careers in biotechnology. Service conditions will be

liberalized for women to be able to return to research/ academics after maternity breaks.

(vii) Creating science & technology leaders for the industry

— The number of Ph.D fellowships offered by the Department of Biotechnology will be increased to 200 per annum

— Public-private partnerships will be encouraged in Ph.D programs through creation of the 'Bio-edu-Grid'- a network of universities and industries facilitating pooling of resources.

— Masters degree level professionals in industry will be encouraged to undertake Ph.D. programs while retaining their jobs through industryuniversity tie-ups.

(viii) Arresting and reversing brain drain

— As mentioned earlier, the number of postdoctoral fellowships offered by the Department of Biotechnology will be increased to 200 per annum in order to attract talent.

— Outstanding young investigator grants in biotechnology will be introduced. This will provide a package including salary support, research grant, equipment and opportunities to attend national and international conferences. The salary support under the scheme will be at par with that of entry-level faculty positions.

— Information on availability of positions in education/ research establishments and industries will be provided on a website to facilitate employment of scientists with specific skills at appropriate positions.

— A database of scientists working in different areas of biotechnology within and outside the country will be created to utilize the expertise appropriately.

(ix) Enabling working conditions for scientists to undertake industry oriented research

— Lateral mobility of scientific personnel: Scientists working at universities and research institutions may be allowed to work in industries for commercialization of their research efforts. This could be in the form of secondment or consultancy with industry or by a sabbatical for three years during the working life of scientists · Dual/adjunct faculty positions: Researchers working in university/research institutions may be allowed to hold positions in the industry and vice-versa

— Joint salary support: Faculty employed in academic institutions may be allowed to hold positions for a period of time in which their salary is contributed both by the industry and the academic institution on a mutually agreed basis. (Such an arrangement will work well only if the teaching requirements of the academic institutions are made obligatory).

— Rapid travel grants: Rapid travel grants scheme for approval within two weeks for young scientists to interact with mentors and industry collaborators would be initiated.

— Institute Innovation grants through the Department of Biotechnology to fund academic researchers to develop their concepts into patentable and more importantly licensable technologies. Such grants may be utilized for the purpose of providing additional infrastructure and manpower, patenting costs as well as costs related to proof of concept studies.

These steps will ensure that the large available resources of human talent in biotechnology are supported and this will guarantee the progress of the biotech sector.

2.2 *Infrastructure Development and Manufacturing*

The strength of a biotechnology company lies in up scaling a number of proven technologies - diagnostics, vaccines, products, and processes - for fine-tuning and large-scale production. While Indian industry is strong in product development and marketing for commercial benefits, biotechnology in India still lacks the infrastructure required to take up R&D in areas like molecular modeling, protein engineering, drug designing, immunological studies, preclinical studies, clinical trials, etc.

In order to get the best from public and privately funded biological/biotechnological research, it is imperative to utilize the infrastructure generated optimally for societal benefits. The concept of contract research organizations (CROs), contract manufacturing organizations (CMOs), contract packagers, lab services providers etc., is steadily taking shape in India. In the area of biomanufacturing, industry estimates that the market for biogenerics in India is expected to see a 43 percent jump from Rs 308.50 crore in 2001 to Rs 1,305.7 crore in 2005 and projected to reach Rs 1,864.3 crore by 2007 registering a growth of 19 percent.

India's strengths lie in the availability of educated and skilled manpower, proficiency in English, low capital and operational costs and the proven track record in meeting international standards of quality. There is ample proof that the Indian companies are committed to global standards. According to reports, outside of the US, India ranks the highest with 61 USFDA-approved plants and in excess of 200 GMP certified pharmaceutical manufacturing facilities.

Strategic actions:

(i) Department of Biotechnology will act to facilitate a Single Window Clearance mechanism for establishing Biotechnology plants.

(ii) Encourage private participation in infrastructure development like roads, water supply and effluent treatment.

(iii) Depositories of biological materials will be created in partnership with industry on IDA model for agriculturally important organisms, medically important organisms, plasmids, cosmids and constructs of special nature generated with adequate human interventions

(iv) State of the art large animal house facilities with GLP will be created for testing candidate vaccines and biotherapeutics. Testing facilities will be created for GMO/LMO

2.3 *Promotion of Industry and Trade*

The emergence of India as a global player in the biotech sector requires government to play the role of a champion and foster an international competitive environment for investment and enterprise development. India's strategy must be to get more value from its R&D investment and from IPR generation. The Biotechnology sector has in recent years witnessed accelerated growth. With approximately 200 industries the growth of the biotech sector in India has been rapid. Current estimates indicate that the industry grew by 39% annually to reach a value of US$ 705 million in 2003-2004.

Total investment also increased in 2003-2004 by 26% to reach US$ 137 million. Exports presently account for 56% of revenue. Currently the biopharma sector occupies the largest market share of 76% followed by bio-agri 8.42%, bioservices 7.70%, industrial products 5.50% and bioinformatics 2.45%. The bioservice sector registered the highest growth (100%) in 2003-2004 with bioagri 63.64% and biopharma 38.55%. The current policy review envisages an annual turnover of US$ 5 billion by 2010. India has to develop its own biotechnological and pharmaceutical products to ensure quality and affordability for global trade. In addition to opportunities in drug discovery and development there are significant openings to provide services to the worldwide biotech and pharmaceutical industries and to leverage low cost high quality manufacturing with a global discovery potential.

Capitalizing on these opportunities would create many new valuable jobs in India as we have seen in the outsourcing and service industry. However, to achieve the targeted business volume, several new challenges have to be met. These are predictable and enabling policies, increased public and private support for early or proof-of-concept stage of product development; improved communication among stakeholders in the sector, public-private partnerships, integration of the Indian biotech sector globally; and improved infrastructure. The vision is to maximize opportunities in the area of contract research, manufacturing and to promote discovery and innovation. The Biotech industry being capital intensive in nature has historically relied on venture capital from public and private sources.

India needs to provide active support through incubator funds, seed funds and provision of various incentives in order to develop the biotech sector. In a highly competitive and fast moving business environment, innovative capacity is an important determinant of the ability to create a continuing pipeline of new products and processes. Innovation covers knowledge creation (R&D), knowledge diffusion (education and training) and knowledge application (commercialization). Innovation is not a one-time event; instead, it has to continuously respond to changing circumstances for creating sustainable growth.

Innovation is measured in terms of external / domestic patent applications; human capital devoted to R&D, government expenditure on R&D proportionate to country's

GDP, business funded expenditure on R&D, indigenous technologies standardized, demonstrated and transferred to industry for commercialization; and the number of spin off companies created. Clear government policies for promotion of innovation and commercialization of knowledge will propel the growth of the biotechnology sector.

Strategic actions:

(i) Innovation:

- Basic and translational research in key biological processes and new materials will be supported as innovation for tomorrow. Access to the knowledge generated will be improved by supporting knowledge and social networks among stakeholders so that those with appropriate skills can convert the research output into useful products and processes.
- Research to promote innovation must be supported increasingly on a cooperative rather than a competitive basis. This requires effective communication among science agencies, research institutions, academia and industry.
- To promote India as a hub of innovation, a network of relevant stakeholders should be developed. Public investment should be used as a catalyst to promote such clustering and networking as this can lead to enhanced creativity by sharing of expertise, resources and infrastructure.
- Availability of human resource would be ensured at each phase of the product cycle.
- Strengthening technology transfer capacity

It is proposed to create several national/regional technology transfer cells (TTC's) over the next 5 years to provide high caliber, specialized and comprehensive technology transfer services. The services would include: evaluating technology and identifying potential commercial uses, developing and executing and intellectual property protection strategies identifying potential licensees and negotiating licenses. Each TTC would service a cluster of institutions in a region or a large city. Optimal delivery of services by the TTCs requires professionals with background in industry and science, wide networks, an external focus and high level licensing skills. The best practices for effective technology transfer will be benchmarked.

The skills of existing technology transfer professionals will be upgraded by a combination of specialized training courses, including linking to important programs redesigning the incentives and career paths for posting. Scientists and other innovators shall be equipped with a better understanding of markets and commercialization pathways, the process of technology transfer, the strategy of protecting intellectual property rights and industrial licensing.

(ii) Fiscal and trade policy initiatives

Biotechnology firms are by far the most research intensive among major industries. On an average the biotechnology sector invests 20-30 % of its operating costs in R&D or technology outsourcing. Government support, fiscal incentives and tax benefits are therefore critical to this sector. These measures will also help to capitalize on the inherent cost effectiveness of the Indian biotech enterprise. The suggested interventions include:

— Exemption of import duties on key R&D, contract manufacturing / clinical trial equipment and duty credit for R&D consumer goods to enable small and medium entrepreneurs to reduce the high capital cost of conducting research.

— Extending the 150 % weighted average tax deduction on R&D expenditure under section 35 (2AB) until 2010 and to permit international patenting costs under this provision and enable eligibility of expenditure incurred with regard to filing patents outside India for weighted deductions u/s 35 (2 ab)

— Enable lending by banks to biotech companies as priority sector lending. Currently banks are almost averse to lending to young biotech companies. In order to encourage banks to lend and provide banking services to the biotech sector, a significant push through appropriate policy guidelines from the Reserve Bank of India is necessary. Currently lending to agri-businesses as well as investment in Venture Funds by banks is categorized as Priority Sector Lending. Biotech as a business has similar characteristics in terms of risk as well as gestation time lines and it is therefore recommended that lending to Biotech be also categorized as Priority Sector lending.

— Remove customs duty on raw materials imported into India, where the finished product is imported duty free. Life Saving Drugs imported and sold in India are exempted from paying customs duty; whereas raw materials for diagnostics and other pharmaceutical biotech products manufactured in India are levied customs duty. To promote the indigenous manufacturing industry and make it competitive globally, raw materials imported by Indian manufacturers should be eligible for Duty Drawback.

— Rationalization of import and export of biological material is considered critical for clinical research and business process outsourcing.

— Simplification and streamlining of procedures for import, clearance and storage of biologicals, land acquisition, obtaining environmental and pollution control approvals would be simplified and streamlined within shorter time frame lines through consultations with various central and state government departments.

- As an effective regulatory mechanism has been put in place though recent interventions, Foreign Investment Promotion Board (FIPB) approval for equity investment may no longer be necessary.
- Joint R&D collaboration and generation of joint IP though global partnerships would be fostered.
- International trade opportunities would be promoted to guide R&D investment Indian biotech strengths would be aggressive by promoted globally.
- Efforts would be made to remove hurdles for contract research especially for input output norms and tax on revenue generated through contract research / R&D.
- Easy access to information, regarding legislation and rules and regulations for transboundary movements of biologicals would be promoted.
- Current standards and safety of products would be enhanced.
- Efforts would be strengthened to promote acceptance of Indian regulatory data internationally.
- Research, trade and industrial partnership would be fostered at regional and sub-regional levels.
- A "cluster" approach would be encouraged to operations. One significant features of the industry is the fluidity and variety of its intercompany relationships, traditionally much greater than in other industries. It has relied to a considerable degree on contracting and outsourcing, especially "upstream" in R&D through various licensing arrangements and "downstream" through co-marketing agreements.
- Collaborative knowledge networks would be promoted. Expanded sharing of information,

including creation/use of collaborative knowledge networks (CKN), can greatly enhance a company's performance under a cluster approach. Managing the many external relationships is complex. Flexible and pervasive communications systems that allow information to flow effortlessly within and between contracting organizations will provide the key to success. Increasingly, IT advances, including webbased approaches, will provide the foundation for these systems.

(iii) Public investment for promotion of innovation and knowledge commercialization

Availability of financial support for early phase of product development to establish proof-of-principle is the key to sustaining innovation. In this context, it is proposed to institute 'Small Business Innovation Research Initiative' (SBIRI) scheme through the Department of Biotechnology in 2005-06 for supporting small and medium size enterprises as a grant/loan. Companies with up to 1000 employees will be eligible. The scheme will support pre-proof of concept, early stage innovative research and provide mentorship and problem solving support in addition to the grant / soft loan. The SBIRI scheme will operate in two phases of innovation and product development.

— *SBIRI Phase - I* :

The funding in this stage will be provided for highly innovative, early stage, pre-proof-of-concept research. Preference will be given to proposals that address important national needs. The maximum amount of funding to an enterprise will be limited to Rs. 50 lakh with not more than 50% of it going as grant and the remaining as an interest free loan. For projects to be considered at this stage, though a partner from a public R&D institution would be considered important, it will not be a mandatory requirement for those companies that have good quality scientists. This should encourage high quality scientists to agree to work in small and medium biotech companies, a change from our traditions. The R & D requirement of the public institution will be met through a grant.

— *SBIRI Phase - II :*

It is expected that some of the proposals funded with SBIRI Phase-I will establish the proof-of-concept. At this stage, the ability of the project to get venture capital funding improves. Such projects will be eligible for Phase-II funding. Some projects could be eligible for direct phase-II support. It is proposed to provide soft loan at this stage for product development and commercialization at an interest rate of 2%. The role of public R&D institution at this stage too is critical. The partner in the public institution at this stage will get the R&D support as grant.

— Small and medium knowledge-based industries in biotech sector will be encouraged to avail of equity support from the SME Growth Fund of Small Industries Development Bank of India (SIDBI).

(iv) Code of best practice for disclosure guidelines

Setting up a 'Code for Best Practice for Disclosure Guidelines for the Indian Biotech Industry'. This Code will be a part of the General Listing Requirements and Disclosure Guidelines and will be in conjunction with SEBI's General Listing Rules and Disclosure Guidelines.

2.4 Biotechnology Parks & Incubators

Establishing biotechnology parks for the growth of the biotechnology industry is essential either through public-private alliance or public/private sponsorships. With its large human resource in molecular biology, microbiology, biochemical engineering, synthetic organic chemistry, chemical engineering and allied branches of engineering and strong institutional base at the universities, CSIR, ICMR and ICAR, India is well placed to support a number of biotech parks. Biotechnology Parks can provide a viable mechanism for licensing new technologies to upcoming biotech companies to start new ventures and to achieve early stage value enhancement of the technology with minimum financial inputs. These biotech parks facilitate the lab to land transfer of the technologies by serving as an impetus for entrepreneurship through partnership among innovators from universities, R&D institutions and industry.

Basic minimum components for parks should include research laboratories for product development, multi-purpose pilot facility for manufacturing and process development, quality control and validation of technologies, common effluent treatment plant, a GLP Animal House, a recognized human resource training centre, administrative support centre etc. The biotech parks should be located so as to be easily accessible for all the stakeholders, tenants, academia with connecting roads, water and power supply and should also attract less administrative clearances from the government.

Strategic Actions:

(i) The Department of Biotechnology will promote and support at least 10 biotech parks by 2010. Each park will necessarily meet the qualifying criteria related to the characteristics of the location, a viable business plan, management strategy and a clear definition of the partners and their roles.

(ii) The Department of Biotechnology will support creation of incubators in biotech parks promoted by a private industry or through public-private partnership in the form of grant upto 30% of the total cost or upto 49% in the form of equity.

(iii) It is proposed that a central body Biotechnology Parks Society of India (BPSI) be set up for the promotion of biotechnology parks in the country on the same lines of the Software technology Parks of India (STPI). The BPSI should be run by professionals having experience in the areas of biotechnology, knowledge in Acts and Rules relevant to biotechnology and management skills. The existing parks can become members of these new biotech parks.

The BPSI would be responsible for evaluating the project proposals and advising the Department of Biotechnology on the funding pattern; facilitating industries in obtaining industrial, environmental and other relevant approvals from the central government; making recommendation regarding fiscal incentives to be granted to the biotechnology parks; providing guidance to the venture capital institutions on investment in biotech parks; providing accreditation to the parks etc.

(iv) Concessions to biotech companies located in biotech parks Biotech companies located at biotech parks are eligible for benefits as per the recent changes in the Foreign Trade Policy:

— Duty free import of equipment, instruments and consumables.

— Tax holiday under Section 10A/ 10B of the Income Tax Act

— A scheme will be put in place for operationalising of the incentives to biotech units located in biotech parks. As a part of this scheme biotech company located in biotech parks to be allowed a five-year time frame to meet the export obligation norms under the SEZ scheme. This measure helps to address the long and unpredictable gestational time lines that are inherent to biotech product development.

2.5 Regulatory Mechanisms

It is important that biotechnology is used for the social benefit of India and for economic development. To fulfill this vision, it has to be ensured that research and application in biotechnology is guided by a process of decision-making that safeguards both human health and the environment with adherence to the highest ethical standards. There is consensus that existing legislation, backed by science based assessment procedures clearly articulates rules and regulations that can efficiently fulfill this vision. Choices are required to be made that reflect an adequate balance between benefit, safety, access and the interest of consumers and farmers. It is also important that biotechnology products that are required for social and economic good are produced speedily and at the lowest cost. A scientific, rigorous, transparent, efficient, predictable, and consistent regulatory mechanism for biosafety evaluation and release system/protocol is an essential for achieving these multiple goals.

Strategic actions:

(i) The recommendation of the Swaminathan Committee on regulation of agri-biotech products and of the Mashelkar committee on recombinant pharma products will be implemented in 2005

(ii) It is recommended that an event that has already undergone extensive biosafety tests should not be treated as a new event if it is in a changed background containing the tested and biosafety evaluated "event". Where adequate evidence is available that the recurrent parent genetic background of a notified/registered genotype is nearly restored (through field data/molecular data), only the agronomic performance and the level and stability of the transgene expression may be analyzed by two-year trial data by the ICAR. Even in case of a structurally altered transgene with no significant modifications in protein conformation, the toxicity and allergenicity tests need not be carried out provided the predicted antigenic epitope remains the same and the level of expression of the transgene is within the defined limits. For the released event, Department is of the view that there is no need of large-scale trials under the Genetic Engineering Approval Committee, as the biosafety aspects have been already addressed adequately before releasing the "event". Only ICAR trials may address the agronomic evaluation of the crop.

(iii) An inter-ministerial group chaired by a reputed scientist will be established in 2005 to address anomalies and issues that arise in regulation from time to time. It is proposed that the administrative support to this committee be through the Department of Biotechnology. The mandate of the committee should be to vet any changes in policies, procedures, protocols by departments dealing with regulation in biotech products and processes; resolve issues emanating from the overlapping/conflicting rules in various acts related to regulation of biotechnology activities in R&D, import, export, releases etc. and to review guidelines, protocols, standard operating procedures and ensure their dissemination to all stakeholders from time to time

(iv) A competent single National Biotechnology Regulatory Authority be established with separate divisions for agriculture products/transgenic crops, pharmaceuticals/drugs and industrial products; and transgenic food/feed and transgenic animal/aqua culture. The authority is to be governed by an independent administrative structure with common chairman. The inter-ministerial group will evolve suitable proposals for consideration of the government.

(v) A centre for in-service training of all professionals, irrespective of their location, engaged in the regulatory process to be established by the Department of Biotechnology in close collaboration with other concerned departments and institutions.

(vi) All existing guidelines to be updated and made consistent with the recommendations of the Swaminathan and Mashelkar committees in 2005. New guidelines on transgenic research and product/process development in animal, aqua culture, food, phyto-pharma and environmental application to be put in place in 2005 by the concerned ministries/departments

(vii) As an interim measure, a special regulatory cell will be created by the DBT to build capacity in the country for scientific risk assessment, monitoring and management, to foster international linkages, support biosafety research; to obtain and review feedback from different stakeholders and provide support to industry and R&D institutions. This cell will only have a promotional and catalytic role

(viii) Measures will be taken to build professionalism and competence in all agencies involved with regulation of biotechnology products

(ix) Research is support of regulation, to safeguard health and environment shall be supported by the concerned funding agencies to generate knowledge that will guide regulations and bioethics policy.

(x) Concerned ministries will make a vigorous effort to promote acceptance of the Indian regulatory decisions by other trading countries.

2.6 Public Communication and Participation

Biotechnology today has become as important as traditional plant and animal breeding have been in the past. At the same time, it raises a number of difficult economic, social, ethical, environmental and political issues that constitute major challenges for the human society. The reception of biotech products by the public has been rather mixed. In general biopharmaceutical products seem to be better accepted than transgenic crops. Clearly it is no longer possible to assume automatic public acceptance of new products and processes that promise public and commercial benefits. Public perception and opinion have a significant influence on the direction and funding of biotechnology research. Hence there is a need to work actively and transparently to inform and engage the civic society in decision-making, and to maintain a relationship of trust and confidence.

The government and the industry must actively promote access to information on the benefits and risks in a balanced manner. To achieve this goal, several enabling factors already exist: a sound biosafety regulatory system; well respected appellate and judicial system for redressal of grievances; cadre of willing and able scientists for effective and accurate communication of information; a large body of extension personnel in agriculture, fisheries, veterinary and human health sectors; large NGO network spread across the country; and an effective and independent mass media. However, several challenges to success need to be recognized while framing the strategies: diverse levels of education and literacy across the country; low understanding of biotechnology among the public; lack of simple communication material; varying quality of science reporting; inadequate inter agency coordination; insufficient dialogue between scientists, industry, policy makers, regulators, consumer for a civil society organizations and the mass media; and lack of sufficiently proactive administrative machinery.

There is a need to build public awareness about opportunities and challenges presented by biotechnology development and to inspire public trust and confidence on the safety, efficacy as well as social and ethical acceptability of products among consumers and civil society through the dissemination of accurate information in a coherent, balanced well articulated, user-friendly and transparent manner. Several focused and well-directed measures are needed to achieve public trust and confidence in biotechnology.

Strategic actions:

(i) Create a cadre of resource persons to reach the stakeholders

— Creation of a cadre of resource persons to provide credible information based on scientific data

— Training media personnel through Institutes of Mass Communication, colleges of journalism and others

— Capacity building among extension personnel in agricultural, fisheries, veterinary and medical sectors

— Involvement of Panchayati Raj institutions in the process of analysis and understanding the risks and benefits associated with GMOs as they will be playing an important role in the local level management of biodiversity, access to benefit sharing etc.

— Awareness generation among undergraduate and post-graduate students in universities, colleges etc on issues related to biosafety.

— Promoting a genetic literacy movement within government and public schools through 50 genome clubs nature clubs each year.

(ii) Creating a media resource network

— To facilitate access to information

(iii) Empowering policy makers

— Regular training programs for policy makers

(iv) Empowering the judiciary

— Setting up a training school for the judiciary under the aegis of Centre for DNA Fingerprinting and Diagnostics, Hyderabad

— Training through the National Law Schools and other similar institutions

(v) Institutional mechanisms for strengthening public trust

— Establishment of a dedicated training centre for biosafety, food and nutrition safety and standards as per codex alimentarius committee.

— Creation of a 'National Biotechnology Awareness Fund' for providing support for the education and preparation of educational resource material for various sections of stakeholders in different regional languages of the country

Section III
Sectoral Road Maps

3.1. Agriculture & Food Biotechnology

Biotechnology is necessary to maintain our agriculture competitive and remunerative and to achieve nutrition security in the face of major challenges such as declining per capita availability of arable land; lower productivity of crops, livestock and fisheries, heavy production losses due to biotic (insects pests, weeds) and abiotic (salinity, drought, alkalinity) stresses; heavy postharvest crop damage and declining availability of water as an agricultural input. Investment in agricultural related biotechnology has resulted in significantly enhanced R&D capability and institutional building over the years. However, progress has been rather slow in converting the research leads into usable products.

Uncertainties regarding IPR management and regulatory requirements, poor understanding of risk assessment and lack of effective management and commercialization strategies have been significant impediments. India owns very few genes of applied value. The majority of the genes under use - about 40 - are currently held by MNCs and have been received under material transfer agreements for R&D purpose without clarity on the potential for commercialization.

The spectrum of biotechnology application in agriculture is very wide and includes generation of improved crops, animals, plants of agro forestry importance; microbes; use of molecular markers to tag genes of interest; accelerating of breeding through marker - assisted selection; fingerprinting of cultivars, land raises, germplasm stocks; DNA based diagnostics for pests / pathogens of crops, farm animals and fish; assessment and monitoring of bio diversity; in vitro mass multiplication of elite planting material; embryo transfer technology for animal breeding; food and feed biotechnology. Plants and animals are being used for the production of therapeutically or industrially useful products, the emphasis being on improving efficiency and lowering the cost of production. However, emphasis should not be on edible vaccines for which use in real life condition is difficult. Nutrition and balanced diet are emerging to be important health promotional strategies.

Biotechnology has a critical role in developing and processing value added products of enhanced nutritive quality and providing tools for ensuring and monitoring food quality and safety. It has been estimated that if Biofertilizers were used to substitute only 25% of chemical fertilizers on just 50% of India's crops the potential would be 2,35,000 MT. Today about 13,000 MT of Biofertilizers are used - only 0.36% of the total fertilizer use. The projected production target by 2011 is roughly around 50,000 MT.

Biopesticides have fared slightly better with 2.5% share of the total pesticide market of 2700 crores and an annual growth rate of 10-15 %. In spite of the obvious advantages, several constraints have limited their wider usage such as products of inconsistent quality, short shelf life, sensitivity to drought, temperature, and agronomic conditions.

From a research perspective the spectrum of organisms studied has been rather narrow and testing has been on limited scale and restricted mainly to agronomic parameters. Environmental factors such as survival in the Rhizosphere / phyllosphere and competition of native microbes have not received sufficient attention. Moreover, results on crops are slow to show. Unless there is a policy initiative at the centre and the state to actively promote Biofertilizers and biopesticides at a faster pace, there is unlikely to be a quantum jump in their consumption. A taskforce headed by Dr MS Swaminthan (2004) under the Ministry of Agriculture has prepared a detailed framework on the application of biotechnology in agriculture. The report rightly lays emphasis on the judicious use of biotechnologies for the economic well being of farm families, food security of the nation, health security of the consumer, protection of the environment, and security of national and international trade in farm commodities.

Guiding principles

Consistent with the overall vision outlined by MS Swaminathan taskforce, the priorities in agri-biotech would be based on social, economic, ecological, ethical, and gender equity issues. The following guiding principles would apply across the sector:

— A comprehensive and integrated view should be developed of r-DNA and non r-DNA based applications of biotechnology with other technological components required for agriculture as a whole

— Use of conventional biotechnologies (e.g. biofertilizers, biopesticides, bioremediation technologies, molecular assisted grading, plant tissue culture etc.) should continue to be encouraged and supported. A precautionary, yet promotional approach should be adopted in employing transgenic R&D activities based on technological feasibility, socio-economic considerations and promotion of trade.

— Public funding should be avoided to research areas of low priority or those that could reduce employment and impinge the livelihood of rural families.

— Regulatory requirement in compliance with Cartagena Protocol, another international treaty and protocol for biosafety, germplasm exchange and access and the guiding principles of codex alimentarius will be implemented through inter ministerial consultative process · Transgenic plants should not be commercialized in crops/commodities where our international trade may be affected. However, their use may be allowed for generation of proof of principle, strictly for R&D, their alternate systems are not available or not suitable.

— In a long term perspective basic research for development of low volume, high value secondary and tertiary products through enabling technologies of genomics, proteomics, engineering of metabolic pathways, RNAi, host pathogen interaction and others. Research and support of biosafety regulation would need support. It is proposed to do away with the large-scale field-testing of the released transgenic events and make it compliant to agronomic test requirements.

Strategic actions:

(i) Accelerating the pace of product development

— In our quest for better products, strong and sustained support should be given to encourage indigenous discovery of new genes and promoters in both public and privately owned institutions. Nevertheless, wherever there be an urgent need for a product to achieve food or nutritional security, creative commercial and academic international partnerships should be explored in national interest for sourcing important genes and promoters through licensing arrangements on exclusive/non-exclusive basis. The cost effectiveness should be carefully assessed on a case-to-case basis.

— A gene bank should be created and be accessible to private and public sector organizations after payment of an appropriate fee.

(ii) Public-Private Partnership

— There is an urgent need to promote and improve the levels of horizontal integration between public-public and public-private laboratories.

— Institutions that generate knowledge and those that specialize in late stage field trials are currently compartmentalized. While support to public-funded innovation must continue to be strengthened, it is proposed that at least 30% of government-funded programmes must have a commercial partner who will be responsible for directing R&D towards commercialization. Public investment should also be encouraged in small and medium companies, especially for late stage trials of transgenic crops. Partnership between public-funded organizations and industry is crucial in the science-to-product chain.

(iii) Inter-ministerial Agriculture Biotechnology Board

— An inter-ministerial Agriculture Biotechnology Board involving Ministry of Agriculture, ICAR, DBT, MoEF, regulatory authority, expert scientists, industry, and the farming community should be established to continuously assess cross cutting issues such as: duplication of R&D investments; capacity building; promotion of horizontal partnerships between various components in the knowledge-product chain; the most cost-effective manner of overcoming nutrition deficiencies (viz. iron, zinc, iodine, vitamin A); availability, access, release and efficient system for biosafety assessment of GMOs and products thereof; safe use of approved technologies and prevention of unauthorized ones; building public trust and understanding biotechnological application relating to global warming, climate change and sea level rise; global trends in consumer/industry preferences of farm commodities. This will also monitor trade and collect market intelligence with respect to GM crops and products and follow the trend of organic markets and watch international developments to identify niche markets, monitor countries that are rejecting GM foods and feed this intelligence to concerned agencies.

Priorities

Priorities for crops and traits should be set after conducting a need assessment exercise in various farming zones. However, an indicative list has been suggested by MS Swaminathan Task Force (2004).

(a) *Crop*: Priority target traits in crop plants would be yield increase, pest and disease resistance, abiotic stress tolerance, enhanced quality, and shelf life, engineering male sterility and development of apomixis. Crops of priority should be rice, wheat, maize, sorghum, pigeon pea, chickpea, moong bean, groundnut, mustard, soybean, cotton, sugarcane, potato, tomato, cole crops, banana, papayas and citrus. In priority crops equal emphasis should be given to GM hybrids and new varieties. The varieties in contrast to hybrids, are preferred by small farmers as they can use their own farm saved seeds for at least three or four years. In case of hybrids, research on the introduction of genetic factors for apomixis would be supported so that resource-poor farmers can derive benefits from hybrid vigour without having to buy expensive seeds every cropping season.

(b) *Livestock*: Priority target traits in livestock would be enhanced fertility and reproductive performance, improved quality, resistance to diseases for reduced drug use, production of therapeutically useful products and quality feed. Livestock of priority would be buffalo, cattle, sheep and goat. Emphasis would be given to animal healthcare, nutrition, development of transgenics and genomics. It is proposed to set up an autonomous institution for animal biotechnology.

(c) *Aquaculture and marine biotechnology*: Application of biotechnology would be crucial in disease resistance, enhanced productivity, fertility and reproductive growth, use of aquatic species as bioreactors for production of industrial products, value added products from sea weeds and other marine taxa and biosensors for pollution monitoring. Species of priority in fisheries would be carps, tiger shrimps and fresh water prawns. It is proposed to set up under the auspices of DBT and autonomous centre for marine biotechnology

(d) *Food and nutrition*: R&D would be focused on: development of biotechnology tools for evaluating food safety, development of rapid diagnostic kits for detection of various food borne pathogens; development of analogical methods for detection of genetically modified foods and products derived there from; development of nutraceuticals / health food supplements/ functional foods for holistic health; development of pre-cooked, ready-to-eat, nutritionally fortified food for school going children; development of suitable pro-biotics for therapeutic purposes and development of bio food additives. It is proposed to set up (under the auspices of Department of Biotechnology) an autonomous institute for nutritional biology and food biotechnology (2006).

(e) *Biofertilizers and biopesticides*: Priorities would include screening of elite strains of micros-organisms and / or productions of super-strains, better understanding of the dynamics of symbiotic nitrogen fixation, process optimization for fermentor - based technologies, improved shelf life, better quality standards, setting up accredited quality control laboratories and standardization of GMP guidelines. Integrated nutrient management system would be further strengthened.

3.2. *Bioresources*

The combined annual global market for the products derived from bioresources is roughly between US$ 500 billion and US$ 800 billion. India is one of the 12 global mega biodiversity centres harbouring approximately 8% of the global biodiversity existing in only 2.4% of the land area. The country is also home to two of the world's 25 hotspots. The varied cultural diversity across the country as well as a very ancient traditional knowledge system associated with the biodiversity represents added assets. Nonetheless, much of this biodiversity is in peril owing, in the main, to anthropogenic causes.

Thus, if the goal of converting our bioresources — animal, plant, microbial and marine — into commercially useful products and processes is to be realized, we need to not only conserve the biodiversity and but also utilize it in a sustainable manner. In this context, absence of a good quantitative information network on bioresources combining remote-sensing data and ground surveys is a major constraint. The situation is even worse for microorganisms. Field- and marine biologists rarely work with molecular scientists and chemists, pharmacologists or other experts, and there is practically no bioprospecting industry. While our traditional knowledge base would be the starting for bioprospecting, ethics and equity should be our guiding principles in benefit sharing.

Animal resources

India is home to an estimated 86,874 species of animals accounting for 7.25% of global animal diversity. The degree of endemism is high and populations of several animal groups are diminishing due to habitat destruction and poaching. Several species, their products and the services rendered by them are crucial to our economic well-being: pollination services by insects (e.g., honey bees, bumble bees, moths, butterflies, beetles, flies) to our agricultural and forestry crops, honey, silk, lac, musk, skins are just a few examples. Other species (e.g., molluscs, frogs, toads, spiders, termites, and snakes) represent potential reservoirs of useful products such as toxins, venoms, enzymes, therapeutic molecules and other bioactive substances. Prospecting for these and other products should be a priority. Biotechnology should be effectively employed for molecular characterization along with bioscreens in search of useful products. Utilization of selected species as bioreactors for production of complex proteins is another important opportunity.

Plant resources

India has a huge treasure of plant resources with over 45,000 known species representing 11% of earth's flora. In terms of flowering plant diversity alone, India ranks tenth in the world. About 33% of flowering plants and 29% of total plants are endemic to the country. Genetic erosion is rampant and conservation is a priority. Prospecting of wild plant resources using molecular approaches and mechanism-based screening should be used to identify novel genes (temperature, drought, salinity tolerant) and gene products (therapeutic compounds, dyes, essential oils, biocontrol agents, gums resins and taxmins). There are potential ornamentals, including foliage - and flower - bearing plants that could be bulked up to be subsequently cultivated on large scale for domestic and international trade. Bioconversion — both cellular and microbial — should be employed to convert intermediates of secondary metabolism into valued added products. Application of genomics, proteomics and metabolomics in carefully selected plants will be very useful.

Biotechnology can contribute substantially in providing cost-effective therapeutically active biomolecules through target/mechanism - based screens, biotransformation, metabolic engineering and transgenic approaches. Biotechnology should also be utilized to add value to our traditional knowledge especially Ayurveda, Sidha and Unani systems as well as tribal and folk medicine. Medicinal plants are also the prime targets of bioprospecting. Besides, the tools of biotechnology can be used for conservation and characterization of plants.

Fossil fuels are chief contributors to urban air pollution and a major source of greenhouse gases (GHGs) - considered to be the main causes behind the climate change phenomena. In contrast, biofuels are renewable; hence, they can supplement hydrocarbon fuels, assist in their conservation, as well as mitigate their adverse effects on the climate. Two major biofuels for the transport sector, bioethanol and biodiesel, are fast becoming popular in many countries around the world. While bio-ethanol (called ethanol) is produced from raw materials such as molasses, beet, sugarcane juice, grains and tubers, biodiesel is produced from oil (derived from oil-bearing seeds such as Jatropha curcas, Pongamia pinnata i.e.karanja).

India imports nearly 70% of its annual crude petroleum requirement. The net oil import bill (import minus exports) was Rs 77,058 crore (Rs 770.58 billion) in 2003-04 as against Rs 74,174 crore (Rs 741.74 billion) the previous year. This expenditure on crude purchase impacts the country's foreign exchange reserves in a big way. The petroleum industry now looks very committed to the use of ethanol as fuel. It is estimated that 75% of the increase in world demand for oil will come from transport. India's transport sector will consume ever-higher amounts of fuel over the coming years. Being one of the largest producers of agro products, including sugarcane, India should take a lead in this worldwide effort at promoting sustainable development.

Microbial resources

Currently only five percent microbes are culturable but there are others of considerable potential value that need to be characterized by new and novel techniques. The five percent culturable microbes have been a source of valuable products. India should play a leading role in the study and utilization of microbial resources. Our priorities include: preparation of inventories based on primary and secondary data; exploration of micro flora in the north-eastern region of the country, and extreme habitats (hydrothermal vents, deep sea sediments, highly acidic, alkaline and anaerobic regions, degraded ecosystems etc.) for discovery of novel bioactive molecules; and study, characterization and screening of uncultivable microbes through appropriate molecular approaches.

Marine resources

The economic zone of the sea as a source of novel genes and gene products - biopolymers, novel enzymes, new therapeutic leads, and other value-added products such as osmo-tolerant crops - has hardly been explored. Marine organisms also present immense potential as biosensors for pollution monitoring as well as bioreactors for production of novel products. Besides, the study of deep-sea organisms including marine microbes has tremendous implications for human health. Expertise in these diverse areas is scattered across a number of agencies/institutions. Strategic Actions would be in the following areas.

Strategic actions:

— There is an acute shortage of expertise in India particularly in taxonomy (the science of the classification of the living and extinct organisms) and microbial ecology. We need to take urgent steps to rectify this.

— Support to capacity building in microbial taxonomy through intensive training programmes at graduate and post-graduate levels

— Promotion of horizontal networking between remote sensing experts, field biologists and computer specialists for inventorisation of bioresources based both on primary and secondary sources of information

— Promotion of closer and effective interaction between biotechnologists, foresters, oceanographers and field biologists.

— Ensure that the use of bioresources be sustainable by regulating the harvesting of medicinal plants

— Formulate a policy to regulate the procurement and sale of medicinal plants in India. Introduce regulatory norms prescribed by DCGI that evaluate the efficacy, safety, and quality of herbal products, which currently are exempt from the scope of any regulation of the DCGI.

— Establish a close working relationship between field scientists, pharmacologists and clinicians so that an all round integration is achieved.

- Public-private partnerships need be promoted for product generation
- Creation of a gene bank for maintaining 'mined' genes
- There is, as on date, only one international depository authority (IDA) in the country at the Microbial Type Culture Collection (MTCC) at IMTECH, Chandigarh; however, for securing our IPR interests, we need to initiate steps to establish a few more centres as IDAs.
- Currently, MTCC does not accept biological materials such as cell lines, cyanobacteria, viruses etc. as it has no expertise or facilities for this purpose. Yet, these are essential for filing patents. IDAs in other countries may refuse to accept such material as they may be potentially hazardous or the shipments may have restrictions. In view of this, the scope of MTCC needs to be expanded by upgrading the existing expertise and infrastructure. Alternately, IDAs should be set up where such expertise and infrastructure are available.
- End products from bioprospecting need to be tested for a variety of parameters before commercial production can begin. There is a need to set up appropriate facilities for such late stage testing of products.
- An autonomous Centre for Marine Biotechnology is proposed to be set up under the auspices of DBT
- An autonomous Institute for Biotechnology for Herbal Medicine under the auspices of DBT is proposed to be established.

3.3. *Environment*

Environmental issues concern everyone. Biotechnology has tremendous potential for application to a wide variety of environmental issues including conservation and characterization of rare or endangered taxa, afforestation and reforestation. It can help in rapid monitoring of environmental pollution, ecorestoration of degraded sites such as mining spoil dumps, treatment of effluents discharged by industries (oil refineries, dyeing and textile units, paper and pulp mills, tanneries, pesticide units etc.), treatment of solid waste, and so on. A number of technologies have already been generated and demonstrated in the country. The real challenge is their adoption by the industry, which has been somewhat uneven. In general, corporate groups have not been overly enthusiastic in adopting biotechnologies even where they have proven efficacy.

The reasons may be several: industry is usually not involved at the planning stage of experiment; enforcement of environmental laws is not always strict or uniform at the ground level and offenders can often escape with impunity; manufacturers frequently change their production schedules based on demand profiles resulting in varied streams of effluents, but microbial consortia specifically designed to one set of effluents may be ineffective in breaking down the changed pollutants.

The goal of environmental biotechnology would be to provide cost-effective and clean alternatives for risk assessment and quality monitoring, eco-restoration of degraded habitats, conversion of toxic recalcitrant chemicals into harmless byproducts, bioremediation of wastes, value-added products from biomass, control of biological invasion through biotechnological interventions, greener process technologies, and effective ex situ conservation strategies. These can be fulfilled through a deeper understanding - and engineering - of the metabolic pathways for degradation of toxicants, environmental genomics and proteomics, and other molecular techniques.

Strategic actions:

For the diffusion of biotechnologies to be successful the following measures should be put in place:

- Ensuring effective and closer horizontal linkages between research workers and the user corporate groups
- Public-private partnership in research and application of clean technologies
- Strict enforcement of the 'polluter pays' principle. This would require interaction with law enforcement agencies
- Capacity building and training, through workshops, of law enforcement officials, municipal workers, state government functionaries and corporate groups on role and relevance of biotechnology in waste treatment
- Steps to encourage small and medium business companies in producing eco-friendly products, microbial consortia etc. for wider usage
- Building greater awareness for protection of proprietary rights of microbial consortia through appropriate methods (e.g., process patent, trade mark etc.)

Greater inter-agency coordination between DBT, MoEF, ICAR, CSIR, CPCB, user agencies and industry through an inter-ministerial Task Force.

3.4. Industrial Biotechnology

At present, a third wave of biotechnology - industrial biotechnology - is strongly developing. Industrial biotechnology (also referred to as white biotechnology) uses biological systems for the production of useful chemical entities. This technology is mainly based on biocatalysis and fermentation technology in combination with recent breakthroughs in the forefront of molecular genetics and metabolic engineering. This new technology has developed into a main contributor to the so-called green chemistry, in which renewable resources such as sugars or vegetable oils are converted into a wide variety of chemical substances such as fine and bulk chemicals, pharmaceuticals, bio-colorants, solvents, bio-plastics, vitamins, food additives, bio-pesticides and bio-fuels such as bio-ethanol and bio-diesel.

The application of industrial biotechnology offers significant ecological advantages. Agricultural crops are used starting raw materials, instead of using fossil resources such as crude oil and gas. This technology consequently has a beneficial effect on greenhouse gas emissions and at the same time supports the agricultural sector producing these raw materials. Industrial biotechnology frequently shows significant performance benefits compared to conventional chemical technology.

Strategic actions:

- Focus in industrial biotechnology will be on reducing chemical and toxic load in our effluent streams, developing non-fossil fuels that are ecofriendly and developing green technologies in Industrial processing.
- Encourage public-private partnership to promote investment in this sector.
- Promotion of industrial biotechnology in strategic areas of manufacturing and developing green technologies.

3.5. *Preventive & Therapeutic Medical Biotechnology*

A healthy population is essential for economic development. Important contributors to the total disease burden are infections like HIV-AIDS, tuberculosis, malaria, respiratory infections and chronic diseases affecting the heart and blood vessels, neuro-psychiatric disorders, diabetes and cancer. It is important to synchronize the technology and products with the local needs of the health system and to facilitate technology diffusion into health practice. Increasing knowledge about pathogen genomes and subtypes, host responses to infectious challenges, molecular determinants of virulence and protective immunity and novel understanding mechanisms underlying escaped immunity and ways to develop novel immunogens will guide development of vaccines against infectious diseases.

Translational research and ability to rapidly evaluate multiple candidates in clinical trials can help accelerate the pace of vaccine development. New directions in manufacturing and delivery are emerging. Major opportunities to control costs are the more efficient processes for manufacturing of new pharmaceuticals, more efficient systems for production of therapeutic proteins and biomaterials and development of drug delivery systems that release drugs at a target site. A shift from parenteral to oral or transcutaneous administration of drugs and vaccine holds the promise of simplifying delivery in health systems.

Medical biotechnology offers a significant possibility for Indian industry to establish a strong pharmacy sector, a growing number of small and medium biotechnology companies, a large network of universities, research institutes, and medical schools and low cost of product evaluation. The medical biotechnology sector annually contributes over 2/3rd of the biotechnology industry turnover. The Indian vaccine industry has highlighted India's potential by emerging as an important source of low cost vaccine for the entire developing world. Further, economic opportunities through contract research and manufacturing through global partnerships are large if supported by enabling government policies and incentives.

The policy goal is to accord high priority to basic and applied research, to strengthen capacity in pre-clinical and clinical product evaluation technologies relevant to all aspects of health and medical care-predictive, preventive, therapeutic and restorative will be supported. Innovation will be supported through new granting mechanisms to support interdisciplinary networks and public private partnerships.

Strategic actions:

(i) Research emphasis

— Basic and applied research would be supported in molecular and cellular biology, genomics, proteomics, system biology, stem cell biology, RNA interference, host response and new platform technologies.

— Pathogenesis of major diseases and molecular mechanics of disease transmission would be investigated

— Product development will be focused on vaccines, diagnostics, new therapies based on cell and tissue replacement, therapeutic antibodies, herbal medicine, plant based medicine, nucleic acids, therapeutics, drug and vaccine delivery systems, new anti microbial agents

— Research to improve production and manufacturing process and local production of biological reagents for development of diagnostics will be supported.

(ii) Improvements in infrastructure and networks

— A centre for translational research will be established. This new institute will be interdisciplinary and will deal with technology policy for public health, molecular pathogenesis of disease, technology development, scale up, product evaluation and technology diffusion into programmes. Centre will be unique in having a pool of scientists, physicians, engineers, and public health persons working on public health grand challenges. This institute will work through public-private partnerships and be a training centre for product development, IPR and regulation.

— A mission mode programme will be initiated in biomaterial and medical device area as an integrated effort by the Department of Science & Technology and Department of Biotechnology. The goal is to promote R&D and industrial activity.

— Two centres of molecular medicine will be supported within medical school system closely interacting with basic science institutes.

— A virtual network of stem cell centres will be established, using a city cluster approach to network scientists and clinicians. Two core stem cell research centres will be established together with several network subclusters. An umbilical cord stem cell bank will be established.

— Mechanism based screening of herbal drugs known in traditional Indian systems would be carried out so as to get value added therapeutics products quickly

— An inter agency task force of ICMR, Department of Biotechnology, and DST will be established to suggest strategies for strengthening medical school based research. Capacity related to translational biology, clinical trials, molecular epidemiology and product development would be strengthened. Integrated MD-PhD programs will be supported.

(iii) Streamline guidelines and procedures for the approval of recombinant pharmaceutical products.

Currently there are multiple regulators, multiple ministries, lack of coordination between these regulators, Over-lapping and duplication of responsibilities of these regulators, lack of a linear progression in the approval process and committees working outside their area of expertise. The Mashelkar Committee (2004) has drawn up a new procedural framework for Biopharmaceuticals, which has streamlined the regulatory process:

— IBSC will monitor all development work (upto 20 Litres) and recommend to RCGM for Animal Toxicity Tests (ATT) & Scale up.

— RCGM will evaluate the recombinant technology & grant permission for scale up - R & D, review and approve for preclinical animal toxicity tests and evaluate ATT data & recommend to DCGI for Human Clinical Trial (HCT).

— DCGI will permit Human Clinical Trials, review Human Clinical Trial Data, grant permission for Manufacture and Marketing the product and inspect the facility where product is manufactured.

— GEAC will review the manufacturing process to ensure that the LMO (living modified organism) is "inactivated" during the process and send its recommendations to the Drugs Controller General of India within the specified time. The GEAC would confine its approval role to LMOs and Category 3 and 4 microorganisms.

3.6. Regenerative & Genomic Medicine

The first wave of real healthy life extension therapies seems likely to result from research stem cells and regenerative medicine which helps natural healing processes to work faster, or uses special materials to regrow missing or damaged tissue. Doctors use regenerative medicine to speed up healing, and to help heal injuries that cannot heal on their own. Regenerative therapies have been demonstrated (in trials or the laboratory) to heal broken bones, bad burns, blindness, deafness, heart damage, nerve damage, Parkinson's and other conditions. Regenerative medicine will result in extended healthy lifespan; we will be able to repair some of the damage caused by aging, organ by organ.

The first crop of simple stem cell therapies for regenerative medicine might be only a few years away from widespread availability. There are major scientific and ethical challenges and safety concerns that must be overcome in taking stem cell based technology for bench to bedside. As it is rapidly evolving field, the existing national (ICMR) guidelines need to be updated and supported by clear articulated procedures. India must consider the potential medical applications of stem cell research. We must reassure end users on the safety and quality by ensuring regulation on stem lines having stable characterizations so that safety risks are predictable. We must reassure suppliers by regulation from lab to market.

Strategic actions:

— Formulate a comprehensive Human Tissue Act (end 2005) with codes and guidance for regenerative medicine. In the intension, ICMR and DBT will support existing guidelines for stem cell research with clear procedures to be followed by scientists and physician.

— DNA and stem cell banking facilities will be created.

— Lay down clearer laws on animal testing in the country for progress to be made in this sector.

— Emphasize on Intellectual Property Rights, confidentiality and feedback

— Regulation for human tissue engineered products.

— Public awareness to be created in order to allay fears through education programmes, industry conferences and seminars.

3.7. Diagnostics for Emerging Medical Paradigm

There is potential to generate a new repertoire of tools for screening people for risk of disease, for early detection of infections and chronic diseases and for predicting outcome. In certain circumstances, single tests are requires to detect multiple pathogens or biochemical abnormalities. To be widely useful, diagnostics need to be real time and low cost.

Advances in biosensors and gene amplification are in the offing to enable real time medicine. Immuno proteomics has the potential to reveal multiple targets for development of diagnostics for diseases for which existing tools are unsatisfactory. For chronic diseases, a shift from treating disease on an individual basis is visualized by genetic assessment of likelihood of benefit from a therapeutic intervention, the so-called personalized medicine. It is seen that most drugs work in only a proportion of patients, targeting therapy to the right sub-group will not only make therapy more efficacious but also make evaluation of newer products cheaper.

Pharmacogenomics is a rapidly growing segment that provides a wealth of information pertaining to defective or missing genes, which call for differentiated medicine - a new avenue for drug research. This emerging discipline combines both infotech and biotech skills in augmenting high-speed data mining of both genotypic and phenotypic information with a view to evolving new forms of medical diagnostics and therapies. Gene regulation and other bio-algorithms will form the core of a new wave of diagnostics that are now being referred to as 'theranostics'. India can be positioned as the hub for differentiated medicine as the country offers one of the most affordable development bases for personalized medicines.

Personalized therapies will demand extensive clinical data generated from well-differentiated patient populations. India has one of the most desired disease and patient profiles that can enable such studies. Coupled with this is the need for a large number of novel diagnostics based on gene and non-gene based platforms. These are clearly large opportunities for Indian Biotech companies to pursue. Personalized drugs also address the affordability factor for expensive therapies such as those that are involved with cancer. Some important barriers to improving the clinical utility of such knowledge exist. These include the highly complex nature of the problem, little incentive for industry to move to genomic-based approach, and lack of provider education.

Strategic actions:

— Establish a cell for Diagnostic Biotechnology to encourage and support studies into the clinical application of pharmacogenomics. This cell should be well positioned to overcome barriers in its work to bring pharmacogenomics to the clinical setting.

— Encourage research-involving investigators with both clinical practice and pharmacology/ pharmacokinetics expertise while at the same time keeping the overall goal of clinical application/utility in focus.

— Provide incentives for this group of clinician-researchers to bring these scientific advances to the patient bedside

— Support education programs to providers of the importance of this field and its utility.

— Encourage biopharmaceutical companies to include pharmacogenomic data in their drug submissions

3.8. Bio-engineering & Nano Biotechnology

Bioengineering covers a wide range of areas such as tissue engineering, biomaterials for therapeutics, biomedical devices and instrumentation, biomedical sensors etc. Tissue engineering, especially of tissues derived from the patient's own cells, offers total acceptance and integration, unlike non-living materials or tissues from other species. Research is focused on developing non-immunogenic materials to serve as scaffolds for regeneration of damaged tissue. Bone and cartilage can be grown today and there is potential for other tissue. Developments in novel biomaterials for micro-particle and nano-particle encapsulated drugs, proteins and other molecules have offered improvement in quality of many therapies with minimal side effects.

Bioengineering offers opportunities for indigenous development of critical implants and extra corporeal devices. Nanoscale structured materials and devices hold a great promise for advanced diagnostics, biosensors, targeted delivery and smart drugs. The application of nanotechnology in bioengineering together with biotechnology offers a great new range of advanced biomaterials with enhanced functionality; and intertwined with tissue engineering, it has the potential to provide true organ replacement technology of the coming decade. While recognizing this potential, it is important to assess not only the efficacy, but also safety of these new interventions regard to human health.

The current market for medicinal devices such as implantables, disposables wound care, dental and orthopaedic materials etc is estimated at around Rs 7000 crores another Rs 5000 crores for the medical instrumentation sector in the country, with a growth rate of 15% per year. Nearly 80% of this demand is met by imports. Major factors limiting the growth of indigenous medical devices industry are the high cost and non-availability of imported technology, higher risks involved in producing and marketing medical devices, inadequate indigenous technology development and production of biomaterials and device and lack of a regulatory authority for medical devices in the country.

Strategic actions:

(i) In bioengineering research emphasis will be on:

— Development of tissue engineered skin, cartilage, cornea, acute liver support, large segment bone repair and small diameter artery

— Biomaterials for drug delivery and controlled release

— Regenerative therapy for the failing myocardium through LVAD support, drug therapy and stem cell technology

— Advanced blood compatible surface fir cardiovascular devices

— Advanced burn and wound dressings

— Bioinstrumentation and physiologic monitoring

— Biosensors for detecting and monitoring metabolites and identifying specific genetic materials and for home monitoring of critical parameters like creatinine, cholesterol and triglycerides

— Dental and orthopaedic materials based on polymer-ceramic composites

— Test methods for safety evaluation of tissue engineered and combinational products.

Pre-eminent applications derived from Nano-biotechnology include drug delivery systems and diagnostics. R&D support will be focused on:

— Micro-electro-mechanical systems (MEMS), medical electronics and fibre optics

— Bio-molecular chips for analysis

— Carbon nanotube based biosensors

— DNA nanowire and electrical characterization of DNA

(ii) Establishing effective institutional mechanisms

— An inter agency working group will be formed to develop a common vision and working strategy in this area

— Appropriate regulatory process will be established to hasten introduction of new medical devices through inter-ministerial consultation

— Focused multi-disciplinary research groups shall be formed with clear mandates, targets and adequate funding; these will be monitored regularly for accountability on research output.

— Suitable institution-industry linkage will be built for technology proving and scaling up of products / medical devices developed at laboratory level

3.9. Bio-informatics and IT - enabled Biotechnology

Bioinformatics has proved to be a powerful tool for advanced research and development in the field of biotechnology. Bioinformatics holds out strong expectations of reducing the cost and time of development of new products such as new drugs and vaccines, plants with specific properties and resistance to pests and diseases, new protein molecules and biological materials and properties. As the full genome sequences, data from micros arrays, proteomics as well as species data at the taxonomic level became available, integration of these databases require sophisticated bioinformatics tools. Organizing these data into suitable databases and developing appropriate software tools for analyzing the same are going to be major challenges. India has the potential to develop such resources at an affordable cost.

Bioinformatics in India can be used effectively for promoting research in biology; prospecting; conservation and management of bioresources; evaluation of products, processes and raw materials, managing complex data required to plan and monitor major national programs; and meeting the growing need of contract services and business outsourcing in pharma and biotechnology sectors. One of the major challenges in optimum exploitation of bioinformatics for solving life science issues is the formulation of appropriate computational biology problems that can be addressed through IT tools. This requires adequate appreciation of the scope and strength of bioinformatics by the biologists and basic understanding of the biological sciences by the information scientists. The solution lies in having adequate leaders with expertise in both life sciences and information technology and strong institutional / program tie-up between specialists from both the fields.

In India, Informatics for life Sciences is an emerging sector - the market size is still quite limited (many verticals each of size USD 20 million - USD 100 million). India has strengths in Chemistry and Computer Science, Software, Health Care and biology. An extensive bioinformatics network has been established covering more than 60 centres spread all over the country. The network has generated human resources through education and training programs at different levels. Some of them have the potential of emerging as advanced R&D facilities. To promote R&D and to utilize the business opportunities would require creation of broadband connectivity, high performance computing facilities, virtual reality centres, availability of high quality trained manpower, interactions with bioinformatics centres in different countries and industry academia interactions for joint database and software creation.

Strategic actions:

(i) Human resource development

- A continuous talent pipeline will be ensured by producing 50-100 quality PhDs, 500 M.Sc and 500 advanced diploma holders in bioinformatics every year
- A national testing program will be put in place for accreditation of students at different levels
- The fellowships of PhD students shall be increased

— Industrial training will be introduced for students pursuing advanced diploma course in bioinformatics

— Virtual classrooms will be established in identified institutions. Teaching material in electronic form will be developed and made available at a reasonable cost.

— Industry participation in developing course content and materials will be ensured.

(ii) Infrastructure development

— Super computing facilities with 10 teraflops computing capacity will be created on biogrid to promote protein folding and drug design activities

— Broadband internet connectivity shall be provided for bioinformatics research and manpower development at subsidized rates

(iii) Testing of public domain resources

— Institutional mechanism will be put in place for testing public domain databases and software and making them available to the users from the academia and the industry. After such testing, these databases and algorithms will be graded so that scientists can use them with higher confidence.

— Commercial databases and software will be tested before the industry invests in the products. Such service will help the industry to reduce their costs and use only certified products

(iv) Inter agency coordination

— There are many government departments and agencies, which are supporting bioinformatics activities. These include CSIR, ICMR, ICAR, DST and MIT. An independent inter departmental agency will be established to coordinate these activities among these departments and agencies.

— The agency will be empowered by legislation to provide the direction and oversee the implementation of the coordinated action plan.

(v) Strengthening of DICs and sub DICs

— The CoEs DICs and sub DICs of BTISnet will be strengthened for hardcore research in bioinformatics as well as high-end human resource development.

— Department of Biotechnology will increase the investment in this sector three times over a period of five years

(vi) Bio IT parks and promotion of bioinformatics industries

— Department of Biotechnology in association with the Ministry of IT will set up bio IT parks for the promotion of the bioinformatics industry.

— High-risk projects in bioinformatics will be promoted through special support mechanism including public-private partnerships.

3.10. Clinical Biotechnology and Research Services

(a) Clinical biotechnology

The cost of launching a new drug into the marked is estimated to cost between $300-500 million of which the cost split between Research and Development is 25% : 75% which would translate to an approximate cost of US$200-400 million for patient clinical studies and trials which form the main components of drug development. The potential of being a key player in this segment is high and remunerative. India has made tremendous progress in clinical biotechnology over the past few years. However, the infrastructure required to identify, document and monitor patients under clinical trials need to be first put in place before India can partake in this activity. There is also an exciting opportunity of conducting longitudinal studies in disease segments for prospecting new biomarkers and novel pharmacogenomic information both yielding high value Intellectual Property.

(b) Research services

With the global pharmaceuticals companies looking outward to reduce their ballooning research costs, a country like India is in a good position to tap the new business opportunities. Due to the emphasis on outsourcing in the nearly stagnant economies to cut costs and retain competitiveness, India is being considered as a destination for contract research in the pharma sector.

Custom research is a services model that most Indian biotech companies have opted for at their start-up stage in order to earn early revenues with which to fund infrastructure and scientist salaries. These companies harbour ambitions of original R&D once they reach a certain profit level. Since research service is about delivery of results, the Government has an important role to play in facilitating the industry. India needs to be promoted as a research service destination. The IP environment is confusing since there is no mechanism or standard for contract sharing. The industry needs to collaborate with the academia and a code of conduct for biotech members has to be designed.

Number of clinical trials that Indian companies would be conducting will increase tremendously for products developed indigenously and imported. Measures are needed that ensure patent safety and compliance with ethical and regulatory requirements. The quality of trials should be such that data generated are accepted globally. Clarity in rules is required related to biotech drugs developed by Indian companies abroad, drugs discovered abroad and licensed to an Indian company, and drugs discovered by an Indian subsidiary of foreign company. The policy goal would be to promote consultations among all stakeholders within the Government and private sector to evolve clear guidelines and procedures.

Strategic actions:

— Frame appropriate rules and procedures to support contract research services through stakeholder consultation.

— Harmonise and streamline the regulatory issues for important and export of biological materials.

— Review eligibility of virtual export of R&D services through contract research for fiscal incentives.

— Address the operational deficiencies through stakeholder consultations for conducting clinical trials.

— Develop a Good Clinical Trial Practice Manual taking into account international guidelines and disseminate these widely.

— Promote, train and support clinical trial investigators as a collaborative ICMR, DBT initiative.

— Strengthen clinical trial capacity in medical schools and hospitals and create centres of excellence.

— Address issues and frame guidelines for patent protection including issue of liability.

— Strengthen institutional ethics committees to bring them at par with global benchmark.

3.11. *Intellectual Property & Patent Law*

The development of capabilities for the effective management of Intellectual Property (IP) is an important element in securing the benefits of public and private sector research in biotechnology. In this context, filings of patents both in India and aboard are critical to the growth of the Indian biotech Sector. The expenses for filing patents especially outside India are prohibitive and a major barrier to effective Intellectual Property Management within the country. Whilst expenses incurred with respect to filing of patents in India is eligible for weighted deduction, similar benefit is not provided for expenses incurred with regard to filing patents outside India. As Intellectual Property Right (IPR)* creation is a pre-requisite for exports to the regulated markets, it is recommended that expenditure incurred with regard to filing patents outside India be also eligible for weighted deduction U/S 35 (2AB). This is also imperative in the new WTO-TRIPS regime, which has taken effect on 1st January 2005.

Strategic actions:

Administration of the new intellectual property rights regime should be improved. This will be achieved by

— Encouraging science graduates to pursue law for better understanding of IPR related issues

— Inclusion of IPR related issues in curriculum of law colleges for facilitating filing of international patents, license negotiation, dispute resolution etc.

— Training scientists and technology transfer professionals in the strategy of intellectual property protection relating to assessment of patentability, prior art examination and technology transfer issues;

— Training patent attorneys on science subject(s) and improving mechanisms for IPR administration through reforms and creation of patent offices, patent codes and ensuring adequate availability of patent attorneys. This will be promoted to an effective inter-ministerial collaboration.

— Setting up of an arbitration council to redress IPR disputes The setting up of an arbitration council will help in improving the perception and increasing International confidence towards IPR protection in India.

— A Rs 50 crore budget be allocated to substantially improve the current Patent infrastructure and set up additional offices in cities such as Bangalore & Hyderabad.

— The Department of Biotechnology will engage in constant dialogue with the Government of India and WTO-TRIPS to address patentability issues in Biotechnology and their future inclusion in the Patents Bill through ammendments.

Section IV
Conclusion

The need for an integrated biotech policy with concurrent attention to education, social mobilization and regulation is considered to be an essential pre-requisite for an orderly progress of the biotech sector. Synergy between technology and public policy is essential for us to achieve an effective mobilization of the tools of new biology for adding both years to life and life to years.

The National Biotech Development Strategy has taken into consideration all the areas that will affect the Indian biotechnology industry. The Policy has clearly chalked out direction to strengthen India's academic and industrial biotech research capabilities, work with business, government and academia to move biotechnology from research to commercialization, foster India's industrial development, inform people about the science, applications, benefits and issues of biotechnology, enhance the teaching and workforce training capabilities and establish India as a preeminent international location for biotechnology.

It is imperative that India leverages resources through partnership and build regional innovation systems. The strategy will help develop local talent for a globally competitive workforce. While it recognizes private sector as a crucial player, the strategy also visualizes government to play a major catalyzing role in promoting biotechnology. The development strategy is based on a strong innovation promotion framework in which industry, academia, civil society organizations and regulatory authorities will communicate in a seamless continuum. The perspective for Indian biotechnology would be global while also concentrating on local issues.

5

Implications of Biotechnology in Trade Communities

Biotechnology is any technological application that uses biological systems, living organisms, or derivatives thereof, to make or modify products or processes for specific use. Scientific and technological developments in the area of biotechnology and genetic engineering have increased significantly in the past decade. These developments cut across a wide range of fields including agriculture, medicine, environment and industry. Compared to any technology in history, these developments are fast paced characterized by polarized debates on the risks and benefits of biotechnology in general and Genetically Modified Organisms (GMOs) in particular.

Biotechnology covers many of the tools and techniques that are commonplace in agriculture and food production. Interpreted in a narrow sense, which considers only the new DNA techniques, molecular biology and reproductive technological applications, the definition covers a range of different technologies such as gene manipulation and gene transfer, DNA typing and cloning of plants and animals. On the other hand, genetically modified organisms (GMOs) are products of biotechnology resulting from genetic engineering techniques and applications. GMOs are organisms that have been genetically engineered.

In simple terms, genetic engineering (GE) is the process of taking genes from one strain of a plant, animal, or virus and inserting them into another, with the goal of reproducing characteristics of the original species in the receiving species. The range of GMOs developed to address problems in agriculture, health and environment has increased. Consequently, this has raised negative concerns and fears. Biotechnology has received far greater acceptance in the field of medicine and healthcare than in the field of food and agriculture. Great skepticism and opposition to transgenic crops in many parts of the world has been witnessed.

The concept of biosafety can be discerned within the expanse framework of the Convention on Biological Diversity. It is under the auspices of the Convention that the Cartagena Protocol on Biosafety was developed and adopted in 2000. Biosafety is concerned with regulatory aspects of biotechnology. It embraces the safe introduction, handling and use of GMOs. The concept of biosafety encompasses a range of measures, policies and procedures for minimizing potential risks that biotechnology may pose to the environment and human health. Establishing credible and effective safeguards for GMOs is critical for maximizing the benefits of biotechnology while minimizing its risks.

Economic theory and classical and neo-classical theories of trade explain trade in terms of the benefits accruing to all the participants engaged in it. The benefits arise from different efficiencies in the production of goods in different countries. Instead of each country producing everything, countries choose to specialize according to their comparative advantages, thereby minimizing costs. Thus trade and specialization are expected to result in higher overall production, with participating economies enjoying a greater share of the expanding pie.

Since trade is theoretically beneficial to everyone involved, the logical conclusion seems to be that trade should be free, and not restricted by any interventions that reduce the realization of these benefits. This theory is well acknowledged in mainstream economics. The on-going debate on biotechnology, biosafety and trade may mark a point of departure from this mainstream theory of trade. The main concern of both theoretical and more practical-political debates is the growing clash between advocates of biotechnology and GMOs in particular seeking unregulated introduction into the world trading system and those who fear its unknown impacts on the environment, human and animal health.

The pervasive nature of biotechnology in the fields of environment, health and agriculture, the speed at which the technology is evolving coupled with the economic interests associated with it, makes it a formidable challenge for both local and international trade. These issues are impacting on the already contested and intricate world trading system. In the recent past there has been growing attention and international discourse on the implications of modern biotechnology on trade.

GMOs in agricultural products have met with consumer resistance both domestically and abroad, resulting in marketing policies that could disrupt trade and international relations. These concerns have generated controversy and are already affecting trade between countries and major trading blocks. Countries such as Australia and New Zealand have established processes for pre-market approval and implementation of mandatory labeling of GMOs.

Regulators in Japan and the European Union have taken a precautionary approach by imposing strict control measures on approval and marketing of GMOs and GM products. They have also imposed mandatory labeling schemes. Since the early 1990s, the European Union has developed and continuously refined a range of legislative measures and regulations to govern GMOs and GM food.

These developments have translated to important concerns of market access for GM products as well as food security.

Implications of GMOs

Pertinent cases related to the implications of GMOs on trade have been documented. With the increasing trend in adoption and commercialization of GM crops, such cases are likely to increase in the years ahead. For instance, three complaints about the EU's restrictive measures affecting GMOs and GM crops were officially brought to the attention of the WTO Dispute Settlement Body by the United States, Canada and Argentina in 2003. These were filed under Article 23 of the Dispute Settlement Understanding (DSU), which states that any WTO Member can initiate a case in the WTO if it considers that its market access rights have been violated.

Britain is reported to have banned the importation of Chinese soy sauce in 2000 because it was produced from soybeans containing GMOs. The experience of the US and EU shows that GMOs can have far reaching implications on trade. The US and EU for many years have been close trading partners representing a typical example of the largest bilateral trading partnerships in the world. For example, in 2002 the value of U.S exports of agricultural products to the EU was $ 6.1 billion, accounting for about 10 percent of all U.S agricultural exports. The main products exported were soybeans, tobacco and animal feed including corn gluten. These commodities are essential in the production of human food and animal feed.

There are also sources of many ingredients used in processed foods. In the same year, the value of EU exports of agricultural products to the U. S. mainly wine and beer was $ 7.9 billion. The US is a leading exporter of agricultural commodities and incidentally the leading producer and exporter of GM crops. This implies that the largest percentages of the crops grown are GM. For instance, in 2003, about 81 percent of soybeans, 73 percent of the cotton, and 40 percent of the corn grown in the U.S was genetically modified. In this regard, the intensification of controversy and public concern over GMOs in the European Union in the 1990s has affected the volume and value of commodities imported from the US. For example, negative consumer attitudes and introduction of new rules that require labeling has led to a drastic decline in the U.S corn exports to the EU. It is estimated that American farmers have been losing about US $ 300 million a year in sale of products such as corn.

In 1998 Europe accounted for only about 4 percent of the total U.S corn exports. In 2002, this had fallen to less than 0.1 percent. Before 1997, the U.S sold about 1.75 million tons of corn annually to Spain and Portugal, the two largest importers of U.S. corn in the EU. But in the 1998-99 crop year, Spain bought less than a tenth of the previous year's amount and Portugal bought none at all. Only one variety of soya bean has been approved in the EU. This has forced US farmers to grow only the soybean variety that has been approved. Consequently U.S soya bean exports to the EU have declined significantly from 9.8 million tones in 1995 to 5.5 million tons in 2002. It should be noted that before 1995 the issue of GMOs had not started affecting trade between EU and US because no GM crop variety had been commercialized.

The first GM crop in the world was commercialized in 1996. Animal feeds such as corn gluten faced less commercial risks compared to foods meant for human consumption. The trade of such animal feeds has not been affected by EU regulations on GM products, but may be greatly affected in the future by the new approval process and stringent thresholds for traceability and labeling if enforced.

Modern Biotechnology

Regulations for introduction, commercialization and trade in GMOs are influenced by a variety of policy frameworks and institutional arrangements at the international, regional and national levels.

International Instruments

There are at least three international frameworks that are concerned directly or indirectly with research into, and trade and use of GM crops. These are the Cartagena Protocol on Biosafety (CPB), the World Trade Organization (WTO) and the Codex Alimentarius Commission. Codex is a voluntary non-binding instrument while CPB and WTO are binding. WTO and Codex came into being before debates on biotechnology gained currency, while the biosafety protocol was specifically developed to govern safe handling transfer and utilization of biotechnology.

The Cartagena Protocol on Biosafety (CPB)

The Biosafety protocol is an international instrument adopted in 2000 under the auspices of the Convention on Biological Diversity. The protocol establishes the international principles that will govern the transfer, handling and use of GMOs with a particular focus on transboundary movement. It is principally concerned with control or mitigation of the risks that may arise from the import of GMOs on the conservation and sustainable use of biodiversity and human health. Although the Biosafety

Protocol is not explicitly intended to be a trade agreement, the fact that its scope includes export and import provisions makes it an implicit or defacto trade agreement associated with the international trade in GMOs. While the biosafety protocol deals with GMOs, WTO was established with the mandate of facilitating free trade between countries by establishing trade rules, serving as a forum for trade negotiations and assisting in the settlement of disputes. WTO deals with trade in all products including GMOs. This happens to be the point of interface between the protocol and WTO.

However, there is confusion and divergence over the relationship between the Protocol and the WTO rules. The protocol and WTO are meant to be mutually supportive, but it then says the Protocol is not to affect the rights and obligations of governments under any existing international agreement. Conversely, it goes on to say that the protocol will not be subordinated to other international agreements. The Biosafety Protocol-controlling trade in GMOs The biosafety protocol covers a wide range of issues to do with international trade in GMOs.

For instance under the Advanced informed Agreement (AIA) which is a key regulatory instrument of the protocol, exporters have to inform countries of their intention to ship a GM product in advance, and receive permission to do so before proceeding. However, only GM seeds intended for planting are covered by the AIA. For the rest of the items, exporters need only to register their shipment with a central Biosafety Clearing House. Governments must then constantly monitor this Clearing House through the internet if they want to be in a position to know what is happening so that they can make informed decisions (acceptance or rejection) an import. This arrangement has generated concerns that many developing countries will not have the capacity to adequately monitor what is going on, (although exporters are obliged to provide a written copy of the information to countries that cannot access the Clearing-House).

The protocol sets procedures for import or export of GMOs intended for direct use as food or feed, or for processing. GMOs that are imported for planting in field trial experiments or commercial planting are subjected to more scrutiny compared to those intended for direct use as food, feed and processing. These will definitely affect the import of animal feeds from countries producing GM feed crops and of commodities from countries producing GM grains. Animals fed on GM feed are not themselves treated as genetically modified. Labeling for such animals or animal products is also elusive because of difficulties in detecting GM content in meat from animals fed on GM grains.

Similarly, there are no mechanisms to detect the use of GM feeds in dairy and poultry products. However, some special markets in the EU and other parts of the world may demand for meat not derived from animals fed on GM feed or treated with GM medications and supplements. The import of GM-grain can be subjected to risk assessment and approval, especially in countries where commodity seed can be used for both food and planting. Importing countries may require that import of grains for food from GM-producing countries be milled at the port of entry and within a certain time, in order to minimize the chance that the seed will be planted and propagated in the country.

World Trade Organization (WTO)

The relationship between WTO and trade in GMOs is an important one. While there is no clear and direct relationship between the two, the current WTO rules under the agreements on Sanitary and Phytosanitary (SPS) Measures and Technical Barriers to Trade (TBT) appear to be applicable to trade in GM products. The SPS agreement governs the use of measures designed to protect animal, plant and human life and health from foreign diseases and pests. The agreement allows members of the WTO to temporarily block trade in the interest of protecting public health based on scientific principles, internationally established guidelines and risk assessment procedures.

The agreement clearly mandates that member countries rely on scientific principles as well as recognized international standards, guidelines, or recommendations when applying food safety regulations in world trade. In this regard, SPS can be applied to regulate trade in GMOs on health grounds. The TBT Agreement governs technical regulations and standards, including packaging and labeling requirements. It requires members to ensure that their national regulations do not unnecessarily restrict international trade.

Codex Alimentarius Commission

The Codex Alimentarius Commission, a joint body of the UN Food and Agriculture Organization (FAO) and the UN World Health Organization (WHO), provides harmonised international food standards, aiming to protect the health of consumers and ensure fair practices in the food trade. The mandate of the commission is to set international food standards that help governments to achieve adequate consumer protection, raise awareness of governments on food safety issues, and help facilitate international trade in foods by preventing unscientific restrictions. Codex Alimentarius Commission is one of the international regulatory frameworks for biotechnology.

Codex standards were first conceived and developed in 1962 but they are voluntary as opposed to being mandatory. In view of this, most countries have adopted them on the basis that they will be more competitive in world trade. Subsequently, national Codex Alimentarius Committees have been established with financial assistance from the FAO in most developing countries. These National Committees comprise of representatives of relevant government ministries, industry and consumer groups. Each national committee sends delegates to international Codex meetings.

Codex standards are explicit as they cover all aspects of food safety including GM products. In this regard, Codex has a number of task forces including the intergovernmental Task Force on Foods Derived From Biotechnology. The ad hoc Intergovernmental Task Force on Foods Derived from Biotechnology was established in 1999 to consider the health and nutritional implications of GM foods. It is mandated to develop standards, guidelines or recommendations, as appropriate, for foods derived from biotechnology or traits introduced into foods by biotechnology.

The task force is expected to submit its final report to the Codex Commission in 2009. The scientific basis for the work of Task Force is provided by FAO and WHO through a series of scientific expert consultations on the safety and nutritional aspects of GM foods. Incidentally, Codex standards are consistent with EU regulations and directives in the sense that they advocate for the need to put in place traceability and food labeling measures. While the Codex standards are only voluntary, the Codex Alimentarius Commission is recognized by the WTO Agreement on Sanitary and Phytosanitary Measures (SPS) as the international organization responsible for standard-setting related to food safety.

The Commission has three standards relating to GM crops:

— Principles for the Risk Analysis of Foods derived from Modern Biotechnology;

— Guidelines for the Conduct of Food Safety Assessment of Foods derived from Recombinant-DNA Plants; and

— Annex on the Assessment of Possible Allergenicity to the Guidelines for the Conduct of Food Safety Assessment of Foods derived from Recombinant-DNA Plants.

Regional Frameworks and Regulatory Instruments

Examples of regulations and policies of regional nature include the EU Directives and Regulations, the Guidelines for Risk Assessment in the ASEAN countries and the OAU Model Law.

European Union regulations and directives

In the last one decade, the European parliament has consistently passed strong laws governing GMOs. This includes a series of stringent legislative measures and stern decision making procedures concerned with handling GMOs from laboratory to field trials. These laws also cover placing on the market and labeling and traceability aspects. The European Union's regulations and directives are the most stringent in the world today. The main pieces of legislation and decision-making procedures concerned with governing of GMOs at various stages (field trials, labeling and traceability and placing on the market) are as follows:

— Directives 90/220/EEC and 2001/18/EC on the deliberate release into the environment of genetically modified organisms (GMOs).

— Regulation 1829/2003 concerning genetically modified food and feed.

— Regulation 1830/2003 concerning traceability and labeling and

Directives 90/220/EEC and 2001/18/EC on the deliberate release into the environment of genetically modified organisms (GMOs).

Directive 90/220/EEC relating to experimental releases and marketing of GMOs entered into force in 1990. Under the provisions of the legislation a number of GM crops including soybean, maize and oilseed rape received authorization. A few years later in 1998, advances in biotechnology prompted member countries of EU to invoke a safeguard clause of Directive 90/220/EEC. The clause allowed member states to temporarily ban a genetically modified product if there was substantial evidence that it implied risks to human health or to the environment. This eventually led to halting of further approvals for introduction of 'GMOs and consequently a de facto moratorium was passed at a EU Environment Ministers Council meeting in June 1999.

After substantial revisions, Directive 90/220/EEC was replaced by Directive 2001/ 18/EC in October 2001. It requires a step-by-step approval process for GMOs. It introduced the following measures to ensure that the regulation of GMOs would meet the demands of EU regulators and EU trading and consumer community. These include:

— principles for environmental risk assessment

— mandatory post-market monitoring requirements, including any long-term effects arising from the interaction with other GMOs and the environment

— mandatory information for the public.

— A requirement for member states to ensure labeling and treaceability at all stages of marketing.

— Commercial approvals for the release of GMOs to be limited to a maximum of ten years

Directive 2001/18 concerns the deliberate release (Part B) and the placing on the market (Part C) of a GM crop, in effect a two-stage process. The first stage called (Part B), covers deliberate release (field trials) of the GM crops. The applicant must submit a dossier of information to the Joint Regulatory Authority, which in the UK consists of experts from the Department for Environment, Food and Rural Affairs (DEFRA), the Advisory Committee for Release into the Environment (ACRE) and English Nature.

If ACRE, and the other official assessors are satisfied that the release poses a very low risk then Ministers are advised that the release be approved for trials. If the field trials demonstrate that the GM crop poses a very low risk, then the applicant can seek to place the product on the market. This second stage is termed Part C. The applicant requests one of the Member States, for example the UK, to forward the dossier to the European Commission in Brussels. The information is circulated to all the Member States and if no member state raises and sustains an objection, then the product can be placed on the market in all Member states.

However, if a Member State raises and sustains an objection, the Commission sends the dossier to a Scientific Committee for assessment. A favourable report will then be sent to a Regulatory Committee for assessment. A favourable report from the Regulatory Committee will result in the Commission adopting the decision and the GM crop can be placed on the market. If the report from the Regulatory Committee is not favourable, the decision can be made by the council of Ministers. The Council of Ministers can either reject or accept the decision on whether or not a product should be placed on the market.

There is no appeal system. If no decision is taken by the Council of Ministers within three months, the decision to accept or reject will be made by the Commission. The product will also have to go through the procedures for authorization of GM food or feed and comply with labeling provisions as required under Regulation 1829/2003. In addition the applicant will have to ensure they meet the traceability requirements as required under Regulation 1830/2003 and maintain these records for a period of five years. On 19 July 2004 under Directive 2001/18/EC the European Commission authorized the placing on the market of Monsanto NK603 GM maize for imports and processing for use in animal feed or for industrial purposes. The decision is valid for 10 years, and imports have to be labeled as containing GM maize.

Regulation 1829/2003/EC on GM food and feed

This regulation was adopted in July 2003. It introduces a centralized authorization procedure for GMOs used as food or animal feed. This means that those intending to market

GM crop in the EU need not to request separate authorizations for the use of the crop as food or feed. A crop is either authorized for both uses, or for neither. The use of GMOs in animal feed did not previously require a specific authorization procedure. The Regulation will thus have an impact on imported GM crops, which are predominantly used as feed for animals.

The regulation is likely to give a considerable advantage to those producers who offer non-GM crops. The Food and Feed Regulation exempts products such as milk and meat, obtained from animals fed on GM crops, from mandatory labeling. With the coming into force of the above new laws and directives, the EU lifted a five year moratorium on genetically modified foods by authorizing importation of biotech maize known as Bt-11 to be sold as tinned sweet corn on supermarket shelves in the European Union.

The authorization was granted in May 2004. However, the lifting of the moratorium does not permit planting of Bt. maize. This is pending and yet to be granted although the EU Health and Consumer Protection authorities contend that GM sweetcorn has been subjected to rigorous risk assessment and has been scientifically proven as being as safe as any conventional maize. Biotechnology companies have expressed discontent about the precautionary approach taken by EU. The existing directives have been described as unfriendly to trade and complying with them makes investment in GM crops and actual commercialization economical unviable.

Regulation 1830/2003/EC on traceability and labeling

Regulation 1830/2003/EC concerning traceability and labeling of genetically modified organisms and treaceability of food and feed products produced from genetically modified organisms came as an amendment to Directive 2001/18/EC which was adopted in 2003. The regulation requires the labeling of all foods produced from GMOs with the exception that food and feed do not have to be labeled if the amount of genetically modified material is below a threshold of 0.9%. With regard to traceability, the regulation requires that GMOs must be traceable throughout the entire production and distribution process. It is one of the most contested directives.

The implications of labeling and traceability of GMOs on trade

Under the new EU legislation, all foods with ingredients derived from GM crops have to be labeled irrespective of whether or not there are detectable levels of GM DNA or proteins. The new regulations also impose for the first time labeling requirements for GM animal feed along the same principles as for GM food. The new legislation does not, however, require labeling of products such as meat, milk or eggs obtained from animals fed on GM feed or treated with GM medicinal products. Products such as cheese or beer, which are often produced with the aid of enzymes produced by GM micro-organisms, also do not need to be labeled.

The rationale for this distinction, according to EU officials, is that these products do not include GM ingredients, but rather are produced with GM processing aids that are not present in the final product. In the EU, it is believed that labeling and traceability are good instruments for restoring consumer confidence in the food regulatory system and GM products. It is seen as the best way to provide consumers with choice about what they eat and restore their confidence by addressing potential environmental or health effects of GMOs.

Proponents in favour of traceability and labeling argue that providing information on food packages will enable individual consumers to choose whether to accept or reject GMOs through their food purchasing decisions. Opponents argue that such labels would unfairly bias consumers against foods that have been determined to be safe to eat by national regulatory authorities. In the light of the foregoing, the issue of labeling has been made either mandatory or voluntary by various countries.

Some governments have enacted legislation making labeling mandatory (e.g. the European Union, Australia, China, Japan, Mexico, New Zealand and the Russian Federation) while other countries reject this approach (e.g. Argentina, Brazil, Canada, South Africa and the United States). Some are considering voluntary labeling for those producers wishing to provide this information to consumers. In countries where labeling has been made mandatory, it comes along with both financial and choice related costs. Labeling implies additional costs that would be borne by food producers and governments initially and could lead to higher food prices and taxes for the consumers. Some of these costs may be imposed on consumers indiscriminately including consumers who don't care about whether the products are GM or non-GM.

Some consumers may be restricted in making food choices by low income or lack of alternative food choices. Labeling may send negative signals isolating GM from non-GM food even where no difference exists between the two as food safety is concerned. Groups opposed to labeling and traceability (mainly U.S government officials, farm groups, biotechnology companies and food processors) argue that such rules are costly, unnecessary and discriminatory against U.S agricultural products. They are likely to disrupt international trade without serving any legitimate food safety or environmental safety objectives. They believe that GM crops and foods are safe as conventional foods and that there is no scientific basis for health or safety restrictions. They point out that labeling is a very expensive affair given that the U.S commodity grain system routinely mixes in GM varieties with conventional varieties of corn and soybeans.

The implications would be that U.S farmers and food producers would have to segregate GM crops and foods derived from such crops at every step of the crop harvesting and competitiveness. Industry and food processing experts also argue that achieving less than 0.9 percent of GM content would be elusive to realize. The US and other countries opposed to labeling and traceability contend that labeling would send negative signals to consumers. Foods derived from biotechnology are likely to be seen by consumers as "warning labels", which would decrease the demand for these products.

For the case of US alone, it is estimated that farmers stand to lose as much as $ 4 billion in annual agricultural exports to the EU as a result of new trade laws and regulations. At the same time, U.S officials argue that EU is exhibiting double standards. EU rules do require labeling of products like beer and cheese (major European agricultural exports that are made using enzymes produced with biotechnology, while soy oil derived from GM soybeans would have to be labeled even if no GM protein could be detected. It is on the strengths of these arguments that the US along with several nations launched a formal complaint process under the WTO's dispute settlement mechanism.

Risk Assessment in the ASEAN Countries

The Association of Southeast Asian Nations (ASEAN) is a ten-member country regional block made up of Brunei Darussalam, Cambodia, Laos, Malaysia, Indonesia, Mynmar, Philippines, Singapore, Thailand and Vietnam. In October 1999 a meeting of the ASEAN Ministers of Agriculture and Forestry was held in Bandar Seri Begawan, Brunei Darussalam. At this meeting, common guidelines on Risk Assessment of Agriculture-Related Genetically Modified Organisms (GMOs) were endorsed. The meeting resolved that:

a) The guidelines would be legally non-binding and would not take precedence over national legislation

b) The guidelines would focus on a science-based risk assessment of agriculturerelated GMOs and would provide a common framework for ASEAN Member Countries to undertake risk assessment of agriculture-related GMOs.

c) Issues such as compensation and liability, labeling, socio-economic and religious factors would not be covered under the Guidelines.

It was also agreed that the objectives of the guidelines would be to provide a common framework for a) Assessment of risks of agriculture-related GMOs to human health and the environment; and b) Scientific basis for decisions relating to the release of agriculture-related GMOs in ASEAN Member Countries. Each ASEAN country is expected to establish the National Authority on Genetic Modification (NAGM) comprising of representatives from national agencies involved in agriculture, trade, economics, environment, health, science and technology and/ or any other agencies and sectors that respective NAGMs may deem appropriate. Apart from acting as a national competent authority in ensuring that introduction and release of GMOs follow national guidelines, the obligation of NAGMs is to ensure the following at the regional level

a) ensure that guidelines are harmonized with regional and international practices

b) establish mechanisms for exchange of information with counterpart NAGMs of ASEAN member countries and other overseas agencies on relevant issues.

c) update the ASEAN Secretariat on any changes in composition of their respective NAGM members

d) inform ASEAN Secretariat of all approvals in the registration of agriculture-related GMOs in their respective countries.

All agriculture-related GMOs brought into the any ASEAN member country by the applicant for release should comply with existing national and international regulations. Before the release of any agriculture-related GMOs in any ASEAN member country, the applicant is required to submit a proposal to the NAGM within whose territory the release is to take place. Once the GMOs under consideration have been granted approval for unconfined or commercial release by the NAGM of the ASEAN country where the application was originally filed, the GMOs shall be registered with the ASEAN Secretariat. The ASEAN Secretariat shall inform the NAGMs of all ASEAN countries and make available the basis for approval.

Apart from the Biosafety Protocol, the African Union (AU) has developed a Model Law for biosafety that can be used as a guide for African countries. The model law does not represent an African wide regional consensus on biosafety. It is a template developed on the understanding that if African countries develop biosafety laws without common elements, such laws may conflict or contradict each other when it comes to dealing with transboundary movement of GMOs. As a result, it is an attempt to develop a mechanism of cooperation to maximize the uniformity of implementation of national biosafety laws.

It is meant to serve as a basis for assisting Member States in drafting their national laws while adapting the provisions of the Model to the realities pertaining to each country. The initiative to develop the Model Law was taken by the OAU and the Ethiopian Environmental Protection Authority. While the objective of harmonizing biosafety legislation is appreciated, the OAU Model Law has been castigated for ignoring existing model legislation in several countries (South Africa, Zimbabwe and Egypt) and draft legislation in the pipeline of adoption in many other African countries. This is likely to lead to diverse and conflicting national biosafety systems. It is argued that in its present form the Model Law is excessively restrictive and will preclude Africa from tapping benefits of the GM revolution.

The Model contains numerous provisions inconsistent with the Cartagena Protocol on Biosafety, which member states have already signed. For instance, the OAU Model includes human pharmaceutical products which had been specifically been excluded from the Protocol. The Model deviates further from the Protocol in that it requires commodities intended for food, feed or processing and their products to be subject to the same extensive biosafety regimes as GMOs. There is no provision for different levels of potential risk.

Biosafety systems in African Countries

African countries have strived to put in place national biosafety systems tailored to the Cartagena Protocol on Biosafety. These efforts are taking place at different levels because of varied capacity scientific and technological capacity in the continent. Only a batch of countries are at an advanced stage (South Africa, Egypt, Kenya and Zimbabwe) while the majority are in the process of catching up. Mid or intermediate level biotechnology countries include Uganda, Tanzania, Zambia, Nigeria, Ghana, Ethiopia and Namibia.

Low-level biotechnology countries include Somalia, Eritrea, Burundi and Sudan. This classification is based on the level of sophiscation of biosafety systems and biotechnology techniques in use ranging from traditional techniques such as fermentation to advanced applications that involve genetic modification. South Africa is leading in the continent with a highly advanced biosafety system comprising of all the key elements of a functioning regulatory system. This includes biotechnology policy, biosafety legislation and biotechnology strategy. In Kenya, a national biotechnology policy and biosafety bill has been developed. They are pending approval by parliament.

A National Biosafety Committee has been created under the National Council for Science and Technology (NCST) to deal with all matters concerning approval of GMOs. Uganda is at par with Kenya in terms of development of policy and legislation for biotechnology but lagging behind in terms of introduction and trials of GM crops. In Tanzania, the National Biotechnology Advisory Council (NBAC) was established to develop the policy within the Tanzania Development Vision of 2025. A draft biotechnology policy has been developed and is being discussed by stakeholders.

In Ethiopia, a country paper on the Status Biotechnology and Biosafety has been prepared. A National Steering Committee on Biotechnology (NSCB) was established under the guidance of the Ethiopian Science and Technology Commission The NSCB developed a draft national policy and strategy on biotechnology that is now awaiting consideration by the government. While these differences exist, by and large, it has been appreciated in the continent that biosafety systems are important mechanisms for ensuring safe use of biotechnology products without imposing unacceptable risk to human health or the environment, or unintended constraints to technology transfer.

While most countries have not developed explicit policies, laws and institutions on biotechnology, they recognize the fact that biotechnology holds a lot of promise in increasing agricultural productivity in a manner that protects the environment. Unlike in the EU, where decisions to do with biosafety are made on a regional basis, the implementation of biosafety provisions and decisions on introduction of GMOs in Africa is currently being done within the policy and institutional framework of individual governments. In this respect, a number of African countries such as Zimbabwe, South Africa and Nigeria have enacted national biosafety legislation by either modifying existing legislation to cover biotechnology issues or created new legislation that deals specifically with GMOs.

Because any new legislation must be in harmony with existing laws, which could be numerous because of the cross-sectoral nature of biotechnology, the configuration of biosafety regulations, policies and institutions slightly varies from one country to another. In Africa, most countries either lack, or are in the process of developing policy frameworks and institutional arrangements to deal with the challenges of biotechnology. While they are slight variations between countries in terms of regulatory policies and institutions, to a large extent the development of biosafety systems has been informed by the requirements of the Biosafety Protocol and technical and financial support from UNEF-GEF project on development and implementation of biosafety frameworks.

Regional networks such as the Eastern Research Network on Biotechnology (BIOEARN) devoted to promotion of biotechnology have been established in the region but efforts to handle biotechnology and biosafety decisions at regional and sub-regional levels are at a nascent stage. The regional and sub-regional trading blocks such as the Common Market for Eastern and Southern Africa (COMESA), Economic Community of West African States (ECOWAS) and Southern African Development Community (SADC) are just beginning to explore possibilities of establishing regional decision making mechanisms and policy for biotechnology and biosafety.

In particular, COMESA and SADC have set in motion a number of activities aimed at pursuing regional collaboration in biosafety. For instance, in 2003 the SADC approved a set of recommendations formulated by the SADC Advisory Committee on Biotechnology and Biosafety as interim measures aimed at guiding the region on issues related to handling food aid; policy and regulations; capacity building and public awareness and participation.

Once countries have functioning national biosafety systems in place, such systems can provide ingredients for common elements and feasible policy options for a regional framework. While regional mechanisms can be complex to design and implement, they can be advantageous in a variety of ways.

The capacity to make informed decisions about GM crops requires adequate funding, human resources and scientific infrastructure. For African countries, the aim should not be necessarily that of single decision-making as in the EU, but of sharing capacity, information and resources. It is possible for the risk management and biosafety communication to be handled collaboratively in a region, but the final decision will of necessity be a national one. A regional approach to biosafey could reduce duplication and improve transparency. They could also offer cost effective options to countries unable to sustain costly biosafety procedures.

While the foregoing merits are appreciated, for reasons of heterogeneity and sovereignty, regional biosafety frameworks should not override or conflict with national priorities as far as choice and decision-making is concerned. An all-encompassing regional framework that meets the expectations of each and every country can be elusive to achieve because of ecological, cultural, political, scientific and technological disparities between and among countries. However, if countries have functioning national biosafety systems, technical and practical elements of such can be harmonized at a regional level. While GMOs have not affected inter-regional trade in the continent, African countries have recently rejected imports of food aid on the grounds of protecting agricultural exports to GM sensitive destinations, in particular the European Union.

For example countries have expressed concerns that allowing into the country GM products can contaminate the conventional varieties and jeopardize the volume and value of commodities exported. It is against this backdrop and vulnerable position that some African countries rejected emergency food aid in the recent past. In 2002, Zambia declined a US offer of maize, some of which contained GMOs. Although, the main issue was that acceptance of GM maize could lead to contamination of local varieties leading to a rejection of Zambian food exports by EC countries, issues regarding the safety of GM maize for human consumption were also raised.

On the other hand, the Zimbabwean Government agreed in July 2002 to allow food aid into the country that contained genetically modified maize, provided it was milled immediately upon arrival to avoid any possible contamination of local varieties. Previously, Zimbabwe had rejected GM food aid due to concerns that it might threaten beef exports to the EU and local maize varieties. Later the Zimbabwean Government agreed to allow GM food aid into the country on condition that it was milled immediately upon arrival to avoid any possible contamination of local varieties.

In Namibia, livestock farmers are concerned that GM animal feed entering the country could threaten the country's meat exports to the EU. The EU is the leading market destination accounting for 80% of Namibia's meat exports. Uganda has exhibited both permissive and precautionary approaches. In 2003, the government of Uganda announced that GMO foods could be permitted for importation into the country but strictly for consumption and not for planting. On precautionary grounds, Uganda has not granted approval for introduction of any GM crops for research trials.

While the foregoing stands out as the position in various African countries, a recent report by the World Bank analyzes the cost and benefits of African countries remaining GM-free. The report indicates that African countries stand to gain more in terms of welfare gains to the society by adopting GMOs than remaining GM-free because of anti-GM policy in European markets. Sub-Saharan Africa countries would not be jeopardized by the developed countries market access barriers for GM crop products.

The magnitude of such consequences will be high in areas such as North America and Argentina which have a heavy dependence on exports of GM sensitive commodities such as maize and oilseeds. In Africa, the concentration of GM sensitive exports and imports is within the continent. This calls for incorporation of biotechnology and biosafety policies in regional and sub-regional trading agreements to ensure that GMOs do not hamper trade in the continent. It is crucial that this is done early enough, although it is likely to take several years before majority of African countries start planting GM crops on a commercial scale.

Attitudes and Perceptions of Public

The extent to which proliferation in GMOs is going to affect trade can be explained partly by public attitudes and perceptions in various parts of the world. This has been characterized by a mix of exuberance on one side and skepticism on the other. On the negative side, opposition is coming from groups concerned about the safety of consuming genetically modified foods, the environmental consequences of growing GMOs and the ethics related to using GMOs. Although few internationally comparable public opinion surveys have been conducted on this issue, the available evidence suggests that public attitudes differ sharply both between and within countries and keep on shifting over time.

In particular, the degree of skepticism and negativity towards GMOs has been high in Europe and less in North America and most of the developing countries. Concerns about GMOs are not about technology illiteracy, it is rather about fear of the unknown. Risk perception analysis shows that high-risk products such as smoking and alcohol are more acceptable to the public than poorly known risks. Studies conducted indicate that attitudes are generally related to income levels, with people from poorer countries having more positive attitudes than those from wealthier countries with a few exceptions. For example, in Asia, higher-income countries such as Japan and the Republic of Korea are more skeptical of the benefits and more concerned about the potential risks associated with biotechnology than people from lower-income countries such as the Philippines and Indonesia.

In Latin America, people in higher income countries such as Argentina and Chile are more skeptical than people from lower-income countries such as the Dominican Republic and Cuba. However, there are exceptions to incomes as a major factor. Within Europe, for example, people from the higher-income countries such as the Netherlands are more positive about biotechnology on average than those from the lower-income countries such as Greece. This shows that other factors such the level of moral and ethical considerations other than income are important determinants.

Also, the potential benefits of biotechnology in terms of reduced use of chemicals and herbicides tend to outweigh the perceived risks in developing countries. In the developed world, such benefits are less convincing against the perceived risks. While the OECD countries, particularly Canada, Mexico and the United States have the highest rate of agreement or positive inclination on growing of and trade in GMOs. Environics International, 2000 conducted the most extensive international study of consumer attitudes towards biotechnology.

Findings from more than 35,000 respondents in 35 countries reveal some clear patterns. For example over two-thirds of respondents from the United States, Colombia,

Cuba, Dominican Republic, China, India, Indonesia, and Thailand agreed that the benefits of GM crops are greater than the risks. On the other hand, fewer than 40 percent of consumers from France, Greece, Italy, Spain, and Japan. Europe, Japan and South Korea saw the benefits as greater than the risks. The stance taken by the European Union with regard to the safety of GM food is attributed to a cultural and historical phenomenon. The health dangers brought about by incidences such as the "mad cow disease," which started in the U.K, in the 1990s, before spreading to the rest of Europe attracted a lot of public attention.

Safety authorities later discovered that the disease could be transmitted to human beings if they ate meat from infected animals. From this experience and other incidences, the Europeans have remained sensitive about food safety and environmental risks that GMOs might pose. A strong movement of opposition to GMOs, which has become deeply entrenched in the EU, has compounded this. It has exacerbated negative perceptions and the level of hostility towards imported (GM) products.

Surveys conducted by have showed that there is more propensity to buy GM foods in the US, while in the EU they have revealed growing skepticism and high level of mistrust associated with biotechnology. Another revelation is that in the EU new directives and regulations are to a large extent driven by public opinion and politics as opposed to scientific evidence. In addition, opposition to GMOs has been heightened by political debates. This is attributed to representation of the Green Party in a number of European parliaments, and the growing number of consumer and environmental groups.

Other reasons against GMOs include information disseminated by environmental pressure groups and strong anti-biotechnology campaigns that have sent warning messages to the public. These organizations have used including mass media alerts, petitions and leaflets as a tool against GMOs. In particular, since the 1990s and 2000s, the media has played a significant part in making GMOs widely known and in highlighting their potential dangers. On the basis of devastating media reports, potential advantages of GMOs have been suppressed or judged as weak and arguments that tried to present the potential advantages of GMOs were often rejected. Another reason for negative perceptions about biotechnology is the perceived role that the technology is seen to be propagating negative globalization, capitalism and economic liberalization.

There is a paucity of information on public perceptions and attitudes towards agricultural biotechnology in the developing countries in general and Africa in particular. This is explained by the fact that very few studies have been conducted on the subject. Attitudes and perceptions in African countries have largely been shaped by Western hopes and fears. In addition statements issued by the government, research institutions, NGOs, academia and the mass media have influenced perception and risks of biotechnology.

One of the most comprehensive surveys in Africa was conducted by the Center for International Development (CID) at Harvard University in co-operation with the South African Labor and Development Research Unit (SALDRU) at the University of Cape Town (UCT) in November 2001. The results of the survey revealed that the significant potential of biotechnology to solve problems of drought and other important agronomic problems such as pest infestation, plant disease, fluctuating yields and high use of pesticides are recognized. Transgenic crops already under cultivation in South Africa, such as pest-resistant Bt cotton and Bt corn, and transgenic crops under development, such as drought-tolerant corn are considered to have a positive economic impact for small-scale and large-scale farmers.

Most respondents agreed that biotechnology was a new tool that helps to solve problems that cannot be solved by conventional breeding, and they strongly disagreed with the statement that genetically modified food poses a serious health risk to consumers. The survey pointed out that the strongest opponents to agricultural biotechnology were representatives of environmental NGOs and the strongest proponents were representatives from scientists representing public and private research institutions. Although South Africa is the most advanced country in biotechnology in Africa and among the top in the world, the response rate to the survey instruments was slow compared to similar studies conducted in Mexico and the Philippines.

There was a very modest response rate from representatives of government institutions and political parties, and the mass media representatives. Most of the target respondents complained about the complexity and technicality of the issues in the questionnaire. Some argued that they did not know much about the risks and benefits of agricultural biotechnology and would be very interested in sources of objective information to give them insights in the first place. Politicians mentioned that they were unfamiliar with the subject. Respondents who were comfortable with the survey were either scientists from Universities (including social science divisions), corporations or government research institutes.

6

Biotechnological Products and Ethical Issues

In the past few years, terrorist attacks have become more daring and have produced greater numbers of casualties than ever before. Many attribute this change to the emergence of a new kind of terrorist, one who believes strongly in extreme violence and who is subject to very different rational constraints than the political terrorists of the past. The 1989 bombing at the World Trade Center in New York and the 1995 sarin attack in Tokyo are often cited as examples of this new brand of terrorism. In these attacks, the main goals were simply to inflict as many casualties as possible upon civilian targets. Fortunately, the attacks did not achieve the widespread damage planned by their perpetrators.

However, a precedent may have been established for enormously destructive terrorist attacks, and the threat of a terrorist use of a weapon of mass destruction is now larger than ever. Biological weapons (BW), in particular, have been singled out as weapons that could be extremely destructive and could potentially be acquired by a terrorist. The Office of Technology Assessment reports that under certain conditions, biological weapons have a destructive potential rivaled only by nuclear weapons. Unlike nuclear weapons, though, the technologies involved in making biological weapons are not particularly difficult to acquire.

Several recent developments may have further eroded some of the constraints that previously existed for obtaining BW. First, the end of the Cold War loosened controls on biological weapons facilities and personnel in the former Soviet Union, providing a source of BW technology that may be relatively accessible. Second, the rapid growth of the biotechnology industry worldwide has hastened the diffusion of technologies that can be adapted for BW development and production. The recent findings in Iraq have highlighted the ease with which states can acquire and develop biological agents. Because biological weapons are so destructive, and because proliferation constraints are eroding, there is a growing concern that they will be used against domestic U. S. targets.

Nature of Biological Weapons

Biological agents are defined as live organisms, or toxins that are derived from live organisms, that are disseminated with the intention of causing disease in the target population. A variety of agents and toxins have been considered for use as biological weapons. The first agent, anthrax, is used in most BW analyses and is often described as the ideal biological agent. It forms hardy spores that can survive in a variety of environments, and when an infective dose of 8,000 to 100,000 sporesis inhaled, it can develop into a virulent disease which is 90% fatal if left untreated.

To provide some evidence of the range of effects that arise solely from the choice of agent, we also consider Venezuelan equine encephalomyelitis virus (VEE virus). VEE virus is considerably more sensitive to the environment than anthrax spores, but its infective dose is only estimated to be 10-100 virions. Also, in contrast to anthrax, naturally contracted VEE is rarely fatal, so it can be viewed as only an incapacitating agent. To be effective a biological agent must somehow be disseminated among the target population.

To disperse BW through the air, one must first generate a cloud of very fine particles, which is then blown downwind and inhaled by the target population. However, the technical aspects of generating such a cloud are not trivial. First, the generated particles must be formed at a very specific size for inhalation and deposition within human lungs. Particles that are larger than tend to deposit in the upper tracts of the respiratory system, where they have a reduced probability of causing an infection.

Particles smaller than are largely exhaled rather than remaining in the lungs. (A is 10meters.) Thus, the optimum size for particle is generally thought to be, although this range can depend upon the specific organism. Second, because biological agents are living organisms, they must be dispersed gently enough to maintain viability. In this regard, anthrax spores are particularly well suited, as they are highly resistant to mechanical stresses and to changes in temperature and humidity. Other agents are more delicate, and much greater care must be exercised when they are dispersed. The combination of these two factors provides a significant technical challenge to any biological weapons designer.

Once in the air, the behavior of particles depends heavily upon their density and size. Anthrax has a density approximately half that of water, and the densities of other bacterial and viral agents are of the same order of magnitude. At this density, particles are transported readily by atmospheric motions and follow all but the smallest fluid motions. Gravitational forces will act to settle them out of the air, but this process occurs quite slowly. (A particle requires approximately five months to settle 500m.)

Other natural processes remove the particles from the air much more efficiently. Raindrops, for example, collect particles as they fall to the ground and clean the air quite effectively. In the absence of precipitation, the main natural mechanism by which these particles are removed from the air occurs as the air motions bring the particles in contact with solid surfaces. When this occurs, the particles tend to adhere to the surface.

Chemical and Biological Weapons

On March 20, 1995, the Japanese apocalyptic religious sect Aum Shinrikyo released the nerve agent sarin in the Tokyo underground system. Thirteen people eventually died and more than 5,500 were injured. This strike was the sect's third intentional and indiscriminate release of sarin within a year. In March 1994, Aum Shinrikyo tried to assassinate the leader of a rival religious sect, the Soka Gakkai, but failed because the spraying system mounted on a van malfunctioned and contaminated its operators.

The second attempt occurred in the town of Matsumoto on June 27, 1994, resulting in seven deaths and 600 injured. While the improved spraying system functioned in this attempt, the targets of the attack-three judges who were expected to rule against the sect in a land dispute-survived with relatively minor injuries as a consequence of a series of errors by the sect's strike team. Following these incidents, terrorism was said to have made a qualitative leap: for the first time a terrorist organization had discharged a weapon of mass destruction.

While some analysts had predicted this development, many still have difficulty understanding the purpose of terrorist organizations resorting to chemical and biological (CB) weapons. Part of this confusion results from the focus on the potential consequences of such an attack: because of their classification as socalled weapons of mass destruction, which lumps them together with nuclear and radiological weapons, CB weapons are said to be able to produce huge numbers of casualties. The immensity of the envisaged consequences defies rational explanation of the political motives for the terrorist attack. Much of the analysis of the threat of terrorism with CB weapons has so far been directed towards circumscribing the threat, by profiling organizations likely to resort to such weapons and investigating the requirements for consequence management.

However, once it has been determined that a particular group has developed an interest in chemical or biological weapons, its eventual acquisition and release of these weapons is virtually taken for granted. Using nuclear weapons as a yardstick, CB weapons are seen as easy and cheap to obtain. This black box approach has diverted attention away from what is actually involved in the acquisition of chemical or biological weapons by a terrorist group.

Previous to the Aum attacks, the best known case of terrorism with CB materials was the indiscriminate use of Salmonella typhimurium, a common cause of food poisoning, by the Rajneesh religious cult in The Dalles, Oregon, in September 1984. The direct motivation was to prevent the inhabitants from re-electing two commissioners of the Wasco County Court who were hostile to the cult's land development plans. The sect had already attracted thousands of homeless people to its community who, because of the liberal voting registration laws, would be able to vote for candidates favored by the sect.

The use of the pathogen was intended to tip the electoral balance further in favor of the sect by incapacitating a large segment of the local population. The cult also attempted to physically harm the commissioners. On August 29, 1984, during a routine fact-finding visit to the Rajneesh community, the two commissioners hostile to the sect were served water contaminated with Salmonella typhimurium. Both judges became sick and one had to be hospitalized. It is unclear whether this act was meant to intimidate or to assassinate the commissioners. Murder was not beyond the pale.

In the planning stages of the plot to decrease voter turnout, sect leader Bhagwan Shree Rajneesh reportedly commented that it was best not to hurt people, but if a few died not to worry. Ultimately, 751 people became ill as a consequence of the restaurant contaminations in September. Despite this apparent success, the cult did not conduct any follow-on attacks; once it realized in October that the plot would fail to alter the election, it gave up the attempt to take over the county. Based on the Aum and Rajneesh cases, as well as other considerations, a distinction is first made between terrorism with CB materials and terrorism with chemical or biological weapons.

The article then deconstructs current views of the terrorist threat from CB weapons, and sketches the evolution of the overall threat from CB weapons since the 1991 Gulf War. Third, it applies the "assimilation model," developed for studying the demand side of CB weapon proliferation in states, to sub-state actors. The assimilation model focuses on the way that political and military imperatives, as constrained by the state's material base, become reconciled with each other so that a new weapon becomes integrated into the state's military doctrine. It can be applied to non-state actors, because it focuses on the many thresholds the promoters of a new armament must overcome and the opportunity costs they are willing to pay to overcome these thresholds.

With nonstate entities, some thresholds identified for states will be virtually nonexistent, while others will feature much more prominently. The presence of certain thresholds and their respective heights can thus provide a way to structure analysis of the armament dynamic for a nonstate actor. Based on these insights, the article concludes that while the acquisition of CB weapons by terrorists is definitely feasible, such organizations nonetheless face enormous obstacles on the path to a CB weapon capability. This decreases the likelihood of their proliferation to terrorists. Moreover, if terrorists acquire such a capability it is highly probable that the quality of the agents will be well below that of similar agents in military arsenals. Finally, from these insights the article suggests ways to prepare for a terrorist strike that utilizes CB weapons.

Terrorism with CB materials deals with the use of any toxic substance or pathogen in pursuit of certain goals. Terrorism with CB weapons refers to the use of warfare agents, that is a toxic chemical designed, developed, and selected by the military to support certain missions laid out in the military doctrine of a state. This distinction highlights the deeper significance of the 1995 sarin attack in the Tokyo underground: for the first time, a terrorist organization turned to a warfare agent. CB terrorism has been practiced throughout history and in all types of civilization.

Poisonous substances, whether animal, vegetable or mineral, have been used for political assassinations or sabotage. Despite the risk of harsh punishments, the prospect of certain success attracted poisoners to the substances. Such use was always limited, however, because only few people had access to the substances and possessed the learning to use them. A qualitative change in knowledge and the accessibility of toxicants took place during the 19th century. With the development and rapid expansion of organic chemistry and the chemical industry, the number of poisonous compounds increased significantly.

Poison appeared in the pre-World War I domestic law of several industrialized countries as part of the penal code or in health, food, drugs, or cosmetics acts. Greater scientific understanding of the propagation of infections contributed to the deliberate use of disease for sabotage. For instance, as part of a program coordinated in Berlin during World War I, German agents cultivated pathogens in the United States and tried to infect horses and livestock ready for shipment to the war theaters in Europe and the Middle East. Chemicals and pathogens were also used in World War II for assassinations and sabotage.

On May 27, 1942, Reinhard Heydrich, Reichsprotektor of Bohemia and Moravia, was allegedly killed by a grenade charged with botulinus toxin supplied by Great Britain to Czech commandos. Soviet agents reportedly had 9-mm pistol bullets containing 22 mg of aconitine for use against German administrative officials in occupied zones. The bullet produced a sure deadly effect even when it failed to hit a vital part of the body. Polish and Soviet partisans were also reported to have used biological agents in sabotage or assassination operations against German troops. Since World War II, poison weapons have been mostly associated with secret services.

In September 1978, the Bulgarian secret police assassinated the exiled writer Georgi Markov with a pellet containing ricin. The toxin is said to have been supplied from the Soviet KGB-run Laboratory 12, which specialized in substances that could kill quickly, quietly, and efficiently. In September 1997, the Israeli secret service Mossad attempted to assassinate the head of the political bureau of the Palestinian militant- Islamic organization Hamas, reportedly with a lethal dose of the synthetic opiate fentonyl. The Truth and Reconciliation Commission produced evidence that South Africa's apartheid regime developed various contraptions charged with a poison or a biological agent for use against the black population as part of its chemical and biological warfare program.

Terrorist organizations, on the whole, have shown relatively little interest in CB materials. Ron Purver's 1995 survey on CB terrorism lists over two dozen reported instances of terrorist use or threat of use of biological materials and a considerable number of threats and incidents with poisonous substances, ranging from apparently empty threats to reports of acquisition and actual discovery of possession. Nevertheless, many of the listed cases could arguably be classified as attempts at homicide, suicide, or criminal extortion motivated by financial rather than political gain. Other cases involved the intelligence services of certain countries.

Common to most examples is the discriminate use of the poisonous agents. Humans were targeted individually; horses and livestock also had to be infected one at a time. Even in those cases in which the assailant is never directly in contact with his victims-e. g., the poisoning with mercury of exported Israeli citrus fruits in 1978 by a Palestinian terrorist organization, or the lacing of foodstuffs in shops with toxicants -the physiological consequences were limited to the person ingesting the toxic substances.

Another shared characteristic is the clear mission-oriented purpose of the attacks with CB materials. In no documented attack with non-warfare agents, whether successful or unsuccessful, were such agents used for their own sake. On the contrary, they were used to achieve an immediate goal, rather than to trigger a large-scale panic by creating a risk of wider exposure to a CB agent. This direct goal-instrument relationship may explain, in part, why no "mass destruction" resulted from these strikes.

The scientific and industrial developments of the 19th century also laid the foundations for chemical and biological warfare in World War I. A huge number of toxic compounds were investigated for their suitability as weapons. In the 20th century, around 70 different chemicals were used or stockpiled as chemical warfare agents. Fewer were standardized, because the selection of an agent represents a compromise:

— A presumptive agent must not only be highly toxic, but also "suitably highly toxic," so that it is not too difficult to handle.

— The substance must be capable of being stored for long periods in containers without degradation and without corroding the packaging material.

— The substance must be relatively resistant to atmospheric water and oxygen so that it does not lose effect when dispersed.

— The agent must also withstand the sheering forces created by the explosion, as well as heat when dispersed.

Thus, for example, the US binary nerve agents were less pure than the unitary ones, but to the proponents of the program in the 1980s the relative ease of production, storage, and transportation, the increased safety for the troops handling the binary munitions, and the less complicated processes of demilitarization and destruction more than compensated for this loss of purity. Moreover, the military had several types of agent at their disposal and, depending on the mission, were able to select them on the basis of volatility versus persistency and lethality versus incapacitation. Candidate biological warfare agents were similarly selected on the grounds of a compromise among pathogenicity, survivability after release, and controllability.

Military biological weapon programs included lethal, incapacitating, and anti-crop agents. This mission-oriented selection of chemical or biological warfare agents resulted in a close goal-instrument relationship. Another common feature of the CB weapon programs was that, especially after World War II, the final production phases (synthesis of the actual warfare agent, manufacture of delivery systems, weaponization, testing) were conducted in facilities owned or controlled by government agencies. This limited the accessibility of these technologies.

Furthermore, the public discourse regarding the necessity of chemical or biological warfare agents in military arsenals changed. As a consequence of the way the military envisaged using these agents, CB weapons were widely viewed as indiscriminate instruments of warfare. The user does not have full control over the agent after release into the atmosphere and, even in a tactical setting, the agent may spread far beyond the primary target area on the battlefield, affecting combatants and non-combatants alike.

The goal-instrument relationship for chemical or biological materials is markedly different from that for chemical or biological warfare agents. This is a direct consequence of the criteria underlying the selection of the agents. The compromises made in the name of military utility may therefore have been a disincentive for terrorist interest in warfare agents. While warfare agents can certainly be used for assassinations or sabotage, there is no immediate rationale available for their selection for these purposes.

Moreover, the terrorist group would have to overcome the many technological difficulties involved in the manufacture, weaponization, and dissemination of these agents. Aum Shinrikyo, of course, did precisely that, but it is the only organization known to have attempted to acquire and use warfare agents on a large scale. The current threat predictions-particularly those involving mass casualties · appear incommensurate with current reality. Before looking into the internal motivations for a terrorist organization to acquire CB weapons, it is therefore necessary to investigate whether the overall threat perception regarding CB weapons has, in fact, changed and, subsequently, been injected into threat projections regarding terrorism.

Part of the problem of rationalizing the use of CB weapons for terrorist purposes lies in the qualification of CB weapons as weapons of mass destruction. This has two major implications. First, it draws the attention of the analyst away from the political motives for resorting to CB weapons and towards the consequences of such employment. As small quantities of toxic chemicals, pathogens, and toxins are said to be able to produce massive casualties, prevention, emergency response, and logistics become the prime focus of policy analysis. The immensity of the envisaged consequences, in turn, defies any rational explanation of the political motives for the terrorist act and adversely affects the assessment of the rationality of the perpetrators.

Second, the grouping of CB weapons with nuclear weapons into the category of weapons of mass destruction blurs the threat and consequence assessments for each individual class of non-conventional weaponry. The most plausible type of weapon to be used in a terrorist strike (chemical weapons) is mentally linked to the most destructive weapon category (nuclear weapons), and vice versa. Chemical weapons are thus implicitly associated with the far greater destructive power of nuclear arms, and the nuclear threat is heightened because of the greater plausibility of terrorist organizations acquiring chemical weapons.

Biological weapons occupy the middle ground between these two extremes: they are said to be both easy to acquire and able to produce mass casualties. Within each of the three categories, the potentially most lethal agents are the ones considered. Furthermore, as Western analysts tend to use nuclear weapons as the yardstick to measure the complexity and cost of armament programs, CB weapons are almost by definition easy and cheap to produce. This, too, affects assessments of the terrorist threat from CB weapons. The focus on the consequences of a terrorist attack with CB weapons has another important implication: it affects a state's "security deficit."

A state always confronts a variety of security challenges. As it can never meet all security contingencies, a security deficit emerges. While the security deficit contains an objective component-for instance, the differences in numbers and types of weapons deployed by two or more adversaries-it is foremost an expression of the subjective appreciation of the threat(s). In the analysis of the terrorism threat, the objective component is by and large absent: new organizations can spring up at different times; their motivations and causes will differ; knowledge of the weaponry at their disposal is fragmentary at best; and the strikes can come without any warning, in any place, and at any time. The only known factors of the security deficit are the state's own vulnerabilities.

Consequently, these vulnerabilities define the threat. The high probability of a terrorist strike with biological weapons is thus assessed on the basis of, for example, the limited understanding of the behavior of pathogens under various environmental circumstances in built-up areas, the presence of essentially unprotected ventilation systems in modern buildings, the limited capability to detect these agents before people are harmed, or the lack of organizational preparedness to respond to the envisaged disaster. In this way, the threat of terrorism with CB weapons rests on worst-case analyses of every conceivable scenario and developments in a wide variety of terrorist organizations, which are then amalgamated into a single threat projection.

Little distinction is consequently made between what is conceivable or possible and what is likely in terms of the terrorist threat. This sense of vulnerability has developed rapidly and its origins are complex. On May 13, 1991, then-President George Bush declared that the United States would forswear the use of chemical weapons "for any reason, including retaliation, against any state" once the Chemical Weapons Convention (CWC) entered into force. The announcement represented a major policy shift. The way in which victory had been achieved against Iraq in 1991 was then seen to have greatly devalued the military utility of CW.

The new weapon technologies had basically rendered chemical weapons obsolete. The confidence of 1991 cannot contrast more starkly with today's extreme sense of vulnerability to CB weapon threats. Several events have contributed to this development. The use of chemical weapons by Iraq against Iranian soldiers in the 1980-1988 war, and against its own Kurdish population, brought the issue of proliferation to the fore. Many companies in Western Europe and the United States had supplied Iraq with key technologies for large-scale production of advanced chemical warfare agents and delivery systems.

The Soviet Union and its satellite states had trained the Iraqi military in the conduct of chemical warfare and sold large quantities of weaponry, some of which Iraqi engineers succeeded in converting into chemical weapon delivery vehicles (e. g., the al-Hussein ballistic missile). At the time, chemical weapon armament programs were also reported in other countries in volatile regions (e. g., Libya and Syria). However, only following Iraq's defeat in the 1991 Gulf War did the world learn of the extent and advanced nature of Iraq's CB weapon programs. Moreover, the great efforts the Iraqi leadership was undertaking to conceal components of these programs from UNSCOM inspectors testified to the high value modern-day proliferators attach to CB weapons.

In addition, in the years following the liberation of Kuwait many soldiers of the coalition forces suffered a variety of medical conditions, collectively known as the Gulf War Syndrome. The lack of conclusive evidence on whether low-level exposure to chemical or biological warfare agents may have been a contributing factor increased the sense of helplessness in the face of such weapons. This sense of helplessness has been further heightened by the possibility that medical pre-treatments to protect soldiers from the effects of CB weapons might actually have caused some of the conditions.

As the events in Kuwait unfolded, the bipolar world order was gradually giving way to a new multipolar international system. Many local and regional conflicts, which had been suppressed during the Cold War, flared up into open wars. Many of the conflicts proved intractable and led to significant casualties for intervening forces. It also gradually dawned on policymakers and military planners that, as a consequence of proliferation, their troops might one day confront an adversary armed with chemical or biological weapons.

Whatever the causes of the Gulf War Syndrome, the phenomenon highlighted many inadequacies in current CB weapon defense, detection, protection, and prophylaxis. For forces unwilling to sustain high casualty rates (especially in view of the remarkably low number of casualties in Kuwait), asymmetrical warfare with CB weapons was suddenly perceived as able to defeat armed forces equipped with the most modern conventional weaponry. Meanwhile, the international community was moving rapidly to strengthen the regimes banning the possession and use of CB weapons.

In January 1993, the CWC was opened for signature. States parties to the 1972 Biological Weapons Convention (BWC) began to consider verification and other measures to significantly strengthen the treaty. However, certain events, beyond the discoveries in Iraq, raised questions about the value of the security offered by these treaties. In 1993, Russian President Boris Yeltsin all but admitted to an offensive Soviet biological weapon program in violation of the BWC. Serious concerns continue to exist about Russia's compliance with the convention.

Trilateral verification and transparency exercises by the three co-depositories of the BWC (Russia, the United Kingdom, and the United States) have come to a halt, feeding suspicions of Russian non-compliance, and, recently, highly publicized accounts by a former ranking official in the Soviet biological weapon program appear to confirm the worst fears. Similar reports have emerged regarding past Russian development of new chemical warfare agents. Neither the agents nor their precursors are featured in the lists of chemicals in the CWC, and may therefore escape detection under its routine reporting and inspection mechanisms.

The rapid deterioration of economic and social conditions in Russia increases the possibility of highly trained specialists with knowledge of chemical or biological weapon development and manufacture being enticed with financial incentives to countries suspected of seeking such weaponry. Low security at the various chemical weapon storage sites in Russia raises the possibility of theft. The disarmament treaties themselves have an impact on the relative threat perception. After the entry into force of the BWC in 1975, CW gradually became seen as the greater threat; in the 1990s biological weapons are once again the larger threat as the CWC sets new standards for verifiability and enforceability. This perception is exacerbated by concerns about poor detection capabilities for biological warfare agents and the problems of consequence management if a release of biological weapons were to occur.

Against the background of the debates on asymmetrical warfare, the CWC ban on in-kind deterrence or retaliation appears to hobble a state party. Yet the whole purpose of disarmament conventions such as the BWC and the CWC is that the parties to them must seek ways of ensuring security by means other than those that are prohibited. This was precisely the deeper sense in President Bush's declaration on May 13, 1991. In a different context, the CWC seems to contribute subtly to the shift in focus towards the consequences of possible chemical weapon employment. The ban on use and preparations for use has removed all possible tactical, strategic, and geopolitical rationales for acquiring chemical weapons from current discussions, leaving only the element of casualty production.

Parallel to this evolution of the CB weapon threat perception, the face of terrorism also changed. A greater number of actors began resorting to such tactics. Terrorist attacks became more lethal, resulting in higher casualty rates per incident and wholesale destruction (although these were entirely due to conventional attacks). Instead of seeking to garner publicity or further a distinct political cause, the new perpetrators of acts of terrorism seem to view the maximization of casualties as a goal in itself. Particularly the religious groups associated with apocalyptic millenarianism, redemptive fanaticism, or racist and ethnic hatred are said to find justification for their acts of violence in the higher authority of God. Because of their belief systems, mass casualties are not an impediment to the furtherance of their goals.

Although so far such groups have mostly carried out their indiscriminate attacks with conventional explosives, they are said to be more likely than other terrorist groups to cross the political and moral barriers to employing CB weapons. Events in the United States have significantly contributed to the new threat perception of terrorism. During the Clinton presidency the United States suffered the first largescale, indiscriminate terrorist strikes on its own territory. The 1993 bombing of the World Trade Center in New York left six dead and around 1,000 injured; the 1995 bombing of the Alfred Murrah Federal Building in Oklahoma City resulted in 168 fatalities and around 500 injured.

Most importantly, the latter attack almost coincided with the release of the sarin nerve agent in Tokyo, creating a mental link between mass casualties and the release of chemical and biological warfare agents by terrorists. A number of factors have thus come together in the 1990s to heighten fears of the likelihood of sub-state actors using CB agents to produce mass casualties. But the prospects of such an event depend on whether terrorists can actually acquire CB weapons, which is a question that has not yet been settled.

Acquisition of CB Weapons

To judge the likelihood of terrorist attacks with chemical or biological weapons a clear understanding of the weapon acquisition process from the perspective of the demand side-the terrorist organization-is required. The demand side is often reduced solely to motivations, such as the relative power and prestige the possession of non-conventional weapons confers to a non-state group, and the difficulties of state retaliation against terrorist groups, because they know no territorial boundaries. Such reasoning is based on a state-level analysis of nuclear weapon proliferation, even though it is far from established that these motivations play any significant role in the acquisition of CB weapons by states.

For all its weaknesses, the 1925 Geneva Protocol, which bans the use in war of CB weapons, eroded the legitimacy of their procurement and possession considerably. Public acknowledgment of such armament programs therefore required extensive justification. Consequently, most possessor states shroud their CB weapon programs in extreme secrecy, so that they cannot assert their relative power and prestige based on these arms. Viewed from the demand side, CB weapon proliferation occurs when a political entity-a state, sub-state, or transnational actor-decides to acquire a CB weapon capability where such a capability does not yet exist, provided this decision is followed by a CB weapon armament dynamic. The armament dynamic that the proliferator must initiate and sustain is the central part of the definition: proliferation is not an automatic process, which, once started, leads to eventual use.

Reversals of the initial decision may occur at any stage as a consequence of, for instance, the impact of dissenting views or insurmountable technical problems. In other words, CB weapon deproliferation occurs as soon as the political commitment to the initial decision ceases to be renewed or if the political entity explicitly reverses that decision (e. g., by unilaterally forswearing the weapons or joining a disarmament treaty). The assimilation model of armament dynamics captures the tension between proliferation and the various pressures towards deproliferation. Assimilation is the process by which political and military imperatives, as constrained by a political entity's material base, become reconciled with each other so that a new weapon, weapon system, or arms category becomes an integral part of the political entity's mainstream military doctrine.

Any weapon, weapon system, or arms category must thus satisfy both political and military imperatives. This implies the existence of a dual decisionmaking track: one on which military appraisals are fundamental, and another on which political considerations play the dominant role. The military track relates to those decisions taken by the military organization to effect a political entity's security policy, including first and foremost the development and implementation of a military doctrine. These planners take into account external factors (e. g., the changing threat) and internal ones (e. g., decision outputs from the political track).

On the political track, overall policy decisions on security and budgetary allocations are taken. As the military and political tracks interact with each other, each decision or set of decisions not only influences future decisions on the same track, but also has ramifications for progress on the other. A considerable level of tension may exist between the tracks, especially if operators on one track make demands that are irreconcilable with the basic goals or premises of actors on the other track. Any initial proposal for a particular type of weaponry envisages a particular end result.

However, the weapon that is actually produced and deployed may differ significantly from the originally anticipated one. This variance between the original concept and the final product is the aggregate of all the opportunity costs paid in the effort to achieve the original concept. The process involves many discrete decisions at the various stages of the armament dynamic. As the proposed weapon enters the decision process, it has to overcome multiple thresholds. These may involve a wide range of issues, including funding requirements, technical difficulties, political opportunism, public opinion, environmental concerns, constraints from international humanitarian law and disarmament treaties, and so on. To overcome such barriers, an opportunity cost must be paid.

It can involve financial expenses as well as the expenditure of political capital to ensure the continuation of the program at a particular stage. Different times and circumstances may thus result in different opportunity costs to be paid for similar decisions in a comparable phase of the armament dynamic. Decisions and conditions hampering the armament dynamic are just as crucial as those promoting it: they affect the outcome by increasing the variance between the original concept and the final product. The nature of the thresholds is determined by intrinsic factors, which relate to the political entity's material base, and extrinsic ones, which refer to the domestic or international environment in which the weapon is conceived.

The political entity's material base constitutes a particularly important independent variable, affecting the decision process on both the political and military tracks. It consists of the political entity's physical base-geographic location, territorial size, population, presence of natural resources, access to resources abroad, etc. -as well as the level of education, of scientific, technological, and industrial development, of economic strength, and so on. It thus involves factors that the decisionmakers can rarely influence within the time frame of the armament dynamic under consideration. In other words, all other factors being equal, differences in the material base of any two political entities may account for different characteristics and results of their respective outputs.

The intrinsic and extrinsic elements may thus raise or lower the opportunity cost for crossing a particular hurdle. Ultimately, should the aggregate opportunity costs be too high for the political entity, the armament dynamic fails, and this is one of the possible causes of deproliferation. The assimilation model views material, political, and societal constraints as obstacles that decisionmakers must overcome if they wish to pursue a particular weapon program, and for which they must be prepared to pay certain opportunity costs. The assimilation model is especially relevant to proliferation because of the extra attention it pays to deficiencies in the material base of the political entity.

Elements (alone or combined) that may play a role in raising a threshold are a scarcity of certain natural resources, a lack of technical skills, an inadequate research or industrial base, and the like. Barring abandonment of the entire project, political leaders may try to develop the missing ingredients indigenously, seek them abroad, or both. Given the probable time frame in which the armament program has to be realized, importing the missing elements may be the only feasible and, in the short run, the cheapest alternative. Especially if the dearth occurs in the physical base of the political entity, importation may be the sole possibility.

Importation of particular technologies, knowledge, or materials is, consequently, one way of structuring the political entity's armament dynamic, albeit one that entails a sizeable opportunity cost. The assimilation model is a heuristic device designed for studying CB weapon armament programs in countries for which limited information is available on decisionmaking processes and the structure of armament programs. The identification of thresholds on the dual decisionmaking track and the assessment of how they may be overcome enable the study of the demand side of proliferation, irrespective of the type of government. The same methodology enables the application of the assimilation model to non-state actors, such as terrorist groups.

The nature of the thresholds can be assumed to be similar for all political entities. However, the relative height of the thresholds will vary among these entities. Certain thresholds identified for states will consequently play only a minimal or no role in a terrorist organization, while other ones will have a far greater relative impact. For instance, the distinct military track on which a political entity formulates its military doctrine is arguably non-existent for terrorist organizations. In this case, from the perspective of the assimilation model, it would mean that all relevant thresholds for the CB weapon armament dynamic in the terrorist organization are located on the political decisionmaking track.

Nevertheless, such a group can be expected to have an idea, however vague, of why it is seeking such weaponry, even if it does not have a fullfledged "military doctrine." The assimilation model, as a heuristic device, suggests that incomplete or imprecise formulation of these goals increases the likelihood of a poor choice of agents, of inadequate dissemination devices and procedures, or of outcomes far below theoretical expectations. The way the political and goal-related (i. e., for a state actor, the military) imperatives are reconciled with each other directly affects the goal-instrument relationship of the selected weapon.

Parameters for Terrorist Organizations

Material Factors

The material base of a terrorist organization seeking chemical or biological weapons is a key determinant, because it consists of elements that the organization can only alter with great investment of resources or time (Figure 1). The physical base, first, comprises elements that determine whether the organization will be materially able to acquire CB weapons. Some elements (e. g., membership size, financial assets, and possession of property and infrastructure) the organization can alter over time through targeted policies. Aum Shinrikyo attempted continuously to expand its membership and to extract the largest possible amount of wealth from its members, its members' families, and its sympathizers. The transfer of property rights, including those of companies, was part of the initiation rites of novices.

A terrorist group has less direct influence over other components of the physical base. Its geographical location and the type of culture in which it is embedded have a direct bearing on the nature of the organization and its success. Aum Shinrikyo enjoyed its greatest success in Japan, where, for example, alienated members of the intellectual stratum of society were receptive to mysticism, and in Russia, where many victims of the social disintegration were similarly seeking solace in various kinds of mysticism. In contrast, the sect was unsuccessful in the United States and Germany, despite some targeted efforts.

Other important components of geographical location for Aum Shinrikyo included the overall level of scientific, technological, and industrial development of Japanese society; the tax exemptions granted to recognized religious organizations, which enabled Aum to amass its considerable assets; and the general hands-off attitude of the Japanese authorities towards religious organizations as a consequence of the religious persecutions before 1945. In other words, the terrorist organization feeds from the society that spawned it.

The second component of the material base consists of the internal characteristics of the terrorist organization. The organization can relatively easily exploit, manipulate, or

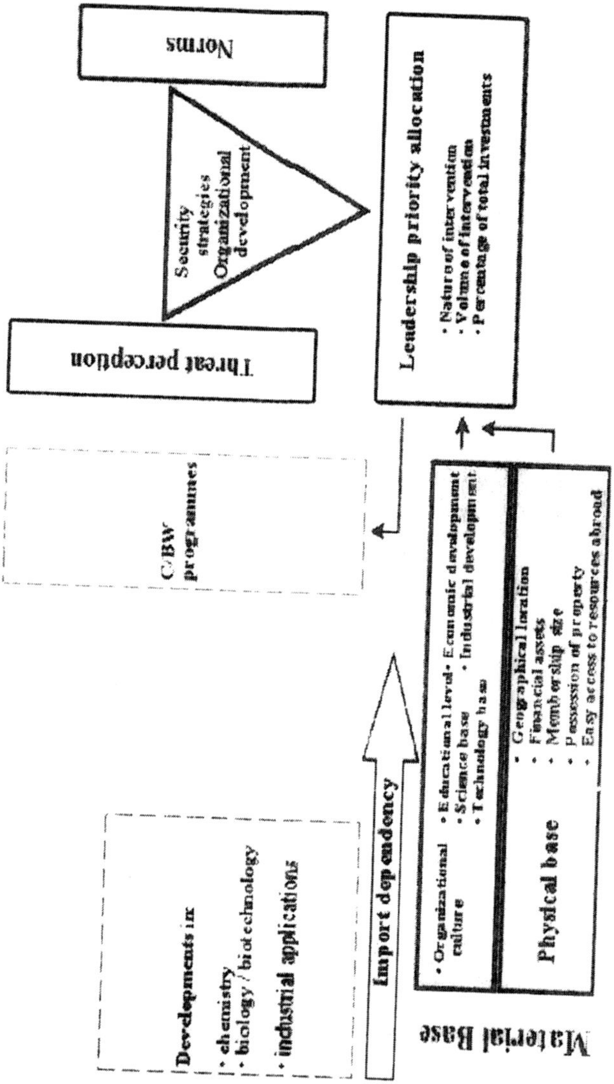

Figutre 1. Terrorist acquisition of CBW key parameters

develop certain of these characteristics. Its culture may be based on social ideology, apocalyptic or millenarian visions, racial superiority, ethnicnationalism, religious fanaticism, and so on. In the quest for chemical or biological weapons the level of education and training of the members as well as the science and technology base they are able to set up become important factors.

Aum Shinrikyo launched repeated recruitment drives to attract promising young scientists and people with other required skills from Japan's leading institutes. These people were able to set up the programs and build the necessary installations. An important weakness in the CB weapon programs, however, was the reliance on relatively unskilled sect members for the operation and maintenance of the installations, which contributed to accidents and leaks. Internal secrecy, and emphasis on dedication to the cause of Aum Shinrikyo in the selection of members to work on the CB weapon programs, were other contributing factors.

Another negative factor on the operational side was Aum Shinrikyo's limited functional specialization. The people in charge of developing the agents were also responsible for developing the dissemination devices and executing the attacks. Their lack of experience in operational planning contributed to the many mistakes and failures. An organization's level of economic and industrial development thus determines the ways it is able to optimize its resources and allocate priorities in advancement of its goals.

Social Environment and Norms

The societal environment in which the terrorist organization evolves provides a second set of factors that influence the leadership in its choices regarding chemical or biological weapons. The tension between the organization's threat perceptions and the internal and external norms that govern its behavior has a major bearing on how the organization will develop and on the security strategies (plans for action or self-protection) it will adopt.

A terrorist group arises as a consequence of the fundamental dissatisfaction of its members with certain (or all) aspects of societal organization. Inevitably, that society will pose a threat to the very existence of the terrorist group. The greater the existential threat to the organization, the greater the chance of its resorting to extreme measures. In fact, this is the shared feature between the Rajneesh cult and Aum Shinrikyo: they both turned to the indiscriminate use of non-conventional means when public authorities threatened the continued functioning of the sects. There is, in addition, a subjective side to the threat perception.

As an officially recognized religious organization, Aum Shinrikyo enjoyed considerable tolerance of its activities by law enforcement agencies, despite many complaints by parents of under-aged sect members and people living near sect compounds, as well as indications of the sect's involvement in murders. Isolation from the broader society was an effective way to hide its CB weaponrelated activities, but also fomented paranoid projections of the threat to the sect posed by Japanese institutions and outside powers like the United States. The risk that the many, rather precise, apocalyptic predictions by the sect's leader, Shoko Asahara, might not be fulfilled provided another incentive to "help" events through chemical or biological weapons.

Norms are another major factor influencing the behavior of the terrorist organization and hence its willingness to pursue CB weapons. Norms, however, form a complex aspect of social interaction and often do not manifest themselves in an absolute, positivist form. For instance, one might anticipate that the release of sarin in the Tokyo underground weakened the norm against the use of chemical weapons or lowered the threshold for other groups to resort to CB weapons. But this would not indicate for whom or in relation to whom the norm was weakened.

The norm against CB weapons has essentially always been one between territorial, sovereign states, that is, between equal partners in the international system.

In view of several gross violations of existing constraints and the unwillingness of the international community to uphold the norm because of ulterior geopolitical interests (e. g., the 1936 war in Abyssinia; the 1980-88 Gulf War; the failure to disarm Iraq despite an explicit UN Security Council resolution) it cannot be said that these norms have been particularly strong. The CWC offers a far stronger norm: not only use, but also possession of chemical weapons and all preparations for offensive chemical warfare are prohibited. The obligations are subject to international verification, and they are enforceable. The CWC obliges each state party to enact domestic penal legislation to ensure that none of its nationals (at home or abroad) or those present on its territory undertake activities in contravention of the treaty.

In other words, since the Tokyo underground attack in March 1995, the norm against chemical weapons has definitely been strengthened and even extended to the sub-state level. Moreover, in the wake of the Aum Shinrikyo attack, Japan has promulgated legislation criminalizing the production, possession, and use of CW. Many other states have also reviewed their existing laws to see whether an event such as the sarin attack is covered or have adopted explicit provisions to that effect in their CWC implementation legislation. The BWC is far weaker in these respects than the CWC, but this norm should also be strengthened once the additional protocol, currently being negotiated in Geneva, enters into force.

In addition, the UN General Assembly adopted the text of the International Convention for the Suppression of Terrorist Bombings on January 9, 1998. Attacks fall within the scope of the convention if they are carried out with an "explosive or other lethal device. " These include not only conventional explosives or other incendiary devices, but also toxic chemicals, biological agents or toxins or similar substances, and radiation or radioactive material. This is the first time that CB weapons are explicitly mentioned in an international counterterrorism agreement.

In short since March 1995 there has been a formal strengthening of the norms against CB weapons for states and sub-state actors and in relation to other states and sub-state actors. However, in practice norms are never absolute and are always weighed against other norms and interests. Domestic enforcement, for instance, may encroach upon freedoms of speech, religion, organization, and so on. It will depend greatly on the maturity of the political and legal system whether a society can differentiate between fundamental rights and criminal activities prepared and executed under the cover of these fundamental rights.

One historical aspect of the development of the taboo against CB weapons, which is often overlooked, is that the civilization that acquired such a mode of warfare clearly understood the military advantage it had over the enemy. It held a monopoly over the surrounding societies. Moral qualms about the application of noxious and poisonous agents in war were not a factor. Consequently, that civilization virtually never formulated legal or moral constraints against these weapons until the monopoly had disappeared or the military advantage had been balanced in an asymmetrical way.

A similar sense of power over the Japanese society, derived from the possession of sarin, was present among the leadership of Aum Shinrikyo. Rather than representing erosion of a taboo, which never existed for the cult, it accelerated the armament process and increased the internal pressures to demonstrate the possession of that power to the outside world. The apocalyptic visions of its leader provided the appropriate social discourse for the new technology within the religious community, but the new technology in turn also helped to determine the group's vision of the apocalypse. The question regarding to whom the norm is applied also hinges on the recognition of the other party as an equal partner.

International norms and laws emerged in the Westphalian state system because sovereign territorial states recognized each other as equal systemic units that could enforce the content of an international agreement within the territory of their jurisdiction. In contrast, a political unit like a religious empire could not and cannot enter into such agreements. First, in such a unit's view sovereignty is derived directly from God and is therefore universal. Consequently, the religious political entity cannot tolerate a separate source of sovereignty.

Second, membership in the entity does not depend on territorial location but on adherence to the faith. The rules, norms, and values of the empire apply to all members of the faith wherever they may be, and do not apply to non-members. Regulations, such as the prohibition of poisoned weapons, governed the conduct of belligerents sharing the same faith, but these weapons were quite permissible against infidels.

History is replete with such examples from all great religions. For terrorist organizations founded in religion, these insights have a double implication. First, the norms maintained by the group may differ significantly from those of the broader society.

Internal or external constraints that could raise the thresholds for acquiring CB weapons on the political track of the assimilation model may therefore simply be non-existent, and the success of the armament dynamic, if undertaken, may depend entirely on factors present in the material base. Second, because of their religious convictions, group members may differentiate themselves from the rest of society to such an extent that the elimination of non-members-even on a large scale-can easily be justified. This worldview may remove any objection to CB weapon use.

Indeed, it may be an important promoter of the armament dynamic in its own right. The strength of norms is also directly linked to the nature of the threat. Sovereign states facing an existential threat or perception that they must meet every security contingency at every level (total reliance on self-help) are less likely to adhere to international norms limiting their options, and are more likely to invest heavily in arms buildups, including chemical or biological weapons, and to defect from international security regimes (like disarmament treaties) if their vital interests are at stake. International law recognizes this tension through, for instance, the inclusion of withdrawal clauses in international treaties.

The International Court of Justice did not contradict this principle principle in its opinion regarding the legality of nuclear weapons of July 8, 1996: it could not conclude whether nuclear weapon use was lawful or unlawful if the survival of the state in question was at stake, despite the potential for massive and indiscriminate destruction. (Hence the smaller surface for "norms" than "threat perception" in Figure 1.). Translated to terrorist organizations, this raises the question of whether an existential threat, especially one that is gradually building up and that the group feels it cannot manage, contributes to the erosion of whatever norms the group might abide by.

Structure of Groups

If the leadership of a terrorist organization decides to embark on CB weapon armament programs, it will have to make some key decisions regarding the allocation of its resources. The decision, and the nature of the program, will depend on the group's security strategies and structure. For example, a loosely structured, amorphous grouping with little central guidance (e. g., many transient right-wing groups and militias in Europe and the United States, including the abortion clinic attackers and the Oklahoma City bombers), or an organization structured in small cells for maximum security, will find it much harder to set up an indigenous CB weapon armament program than a vertically integrated and ideologi-cally uniform group, such as Aum Shinrikyo or the Rajneesh sect.

Organizations are also constrained by their material bases and may have to seek many, if not all, ingredients and technologies from outside. The nature and size of these constraints determine the degree to which a group must rely on external sources for its technologies, commodities, and expertise. For a terrorist organization this can be a formidable challenge. Unlike a state actor, which can buy technologies abroad and hire specialists, a terrorist organization must work in total secrecy because of the absence of a safe haven on the territory it occupies and the constant threat that law enforcement officials may raid the facilities. This means, for example, that the organization cannot hire a specialist or technician for a limited time to solve a certain problem, but must recruit him and convince him of the justness of its cause.

This import dependency is also a function of the relative complexity of the weapon system the leadership has decided to acquire. With the key components in place, the armament dynamic can continue along the dual track until the desired weapon is achieved. Decisionmakers must overcome the various thresholds and pay the various opportunity costs, while trying to keep the variance between the envisioned and the actual weapon as small as possible. The actual chemical or biological weapon in the hands of the terrorist organization will reflect the aggregate opportunity cost paid along the way (e. g., in terms of the quality of the agent): if the aggregate opportunity cost is too high for the organization, then the armament dynamic has failed (e. g., Aum Shinrikyo's botulinus toxin and anthrax programs).

The influence of the various parameters can be illustrated when comparing Aum Shinrikyo with the Rajneesh sect. The Rajneesh sect was responding to a rapidly evolving crisis that threatened its continued existence. The person in charge was a qualified nurse with sufficient skills to cultivate a pathogen, but not to set up a sophisticated biological weapon program. Moreover, the cult had no time to develop its material base. The goal was limited in scope and time, namely influencing the outcome of local elections. The sect thus opted for an incapacitating rather than a lethal agent, thereby narrowing the technical requirements for the laboratory.

The choice of a Salmonella strain, which causes food poisoning, also simplified dissemination, as a liquid solution could be poured on food in public places. In addition, this reduced the need for functional specialization in the sect. The straightforward goal-instrument relationship also meant that as soon as the sect realized that it would not attain the desired outcome, it terminated its program. Aum Shinrikyo's plans were far more ambitious: it sought to destabilize Japan and eventually take over all its governmental functions. To this end, the sect pursued a broad set of instruments, including conventional weapons, an earthquake machine, a laser gun, and a nuclear device, as well as CB weapons.

While many accounts of Aum Shinrikyo's activities have focused narrowly on the CB weapon programs, the important point for demand- side proliferation analysis is that the sect actively sought a broad range of weaponry. This had two major implications. First, the element of priority resource allocation by the sect leadership became important in the CB weapon armament dynamic. The sect spread its huge financial assets and other resources over several weapon programs, as it tried to become self-sufficient in every area.

It even opted to establish its own production line for the Kalashnikov AK-74 rifle instead of purchasing the required firearms, even though the black market in Russia, where Aum Shinrikyo had a large following and many regional centers, could have provided ample opportunities. Each program placed increasing demands on manpower, the ability of the offices outside Japan to purchase the required technologies, and so on. Moreover, each program created its own follow-on imperatives. The prospect of mass-produced assault rifles, for example, raised the issue of training sect members to use these weapons and, in turn, placed fresh demands on the recruitment drive (e. g., to attract highly trained military personnel as instructors).

Had the sect concentrated its resources more on the CB weapon programs, it might have achieved greater success in terms of creating a viable biological weapon or larger production batches of higher-quality chemical warfare agents. As it turned out, the sect was successful in few of its weapon programs. Second, there was no rationale for the CB weapon programs without the other weapon programs. Aum Shinrikyo's ultimate goals were the creation of Armageddon, the toppling of the Japanese government, the subjugation of the Japanese population, and, finally, the establishment of its own form of governance.

CB weapons could conceivably play a role (e. g., through the creation of mass panic and exposing the weaknesses of the authorities to protect the population) in the first three phases, but were insufficient in and of themselves. Any large-scale release of chemical or biological warfare agents in isolation would invite a massive response from the law enforcement authorities (as ultimately happened after the Tokyo underground attack), leading to the potential demise of the organization. In other words, it was impossible in practice for Aum Shinrikyo to concentrate its resources on CB weapons.

In view of its grand strategy, the leadership had to spread its large, but nonetheless limited, resources over the various programs. From the perspective of the CB weapon programs, this imperative raised the thresholds in the assimilation process, which contributed to the reduced quality and quantities of the chemical warfare agents and to failures with respect to the biological warfare agents. In summary, the factors that contributed to the establishment of the CB weapon programs were ultimately also responsible for the rather poor results.

7

Governing Biosafety in India

Increased use of biotechnology in sectors such as agriculture and medicine has been accompanied by conflict over the ecological, socioeconomic, political and ethical implications of such use. A central reason for conflict has been that biosafety governance has to be largely anticipatory, given scientific and normative uncertainties surrounding risks and benefits associated with dissemination and use of this technology.

Transnational Governance

The Cartagena Protocol on Biosafety, negotiated under the Convention on Biological Diversity, mandates the "advance informed agreement" of an importing country prior to the transboundary transfer of certain GMOs (called "living modified organisms" or LMOs here). The objective is to allow an LMO receiving country to assess potential risks to biological diversity and human health that could be posed by such transfers. The demand for a biosafety protocol and for "informed consent" prior to GMO trade came originally from developing countries, led by Malaysia, during negotiation of the Convention on Biological Diversity in the early 1990s.

While this developing country demand was supported by green groups and Nordic countries like Denmark, it was opposed by agricultural GMO producer and exporter countries such as the United States and Australia, as well as by biotechnology industry groups. Developing countries called for a biosafety protocol out of concern that they might become the testing grounds for what they perceived to be novel substances which they did not have the capacity to deal with. However, those opposed to a protocol argued that GMOs did not pose risks different from those associated with other techniques of genetic manipulation such as traditional breeding, and hence did not merit separate international regulation.

Although the European Union initially offered only lukewarm support for a biosafety protocol under the CBD, negotiations of the protocol have unfolded over a four-year period of expanding public concern in Europe over ecological and food safety concerns relating to genetically modified organisms. This has been accompanied by an escalating trade conflict between the United States and the European Union. A de facto moratorium has been in effect against entry of transgenic crops into Europe over the last two years, as the European Community has debated amendments to its regional directives on contained use and deliberate release of GMOs and has halted new approvals until such amendments are in place. This has transformed what began largely as a developed versus developing country issue into a growing intra-OECD conflict, with repercussions for the scope and clarity of the protocol's obligations.

The evolving nature of the alliances in this negotiation is reflected in the fact that towards the end of the failed Cartagena meeting in February 1999, five distinct negotiating groups had emerged. These included the Miami Group, consisting of six agricultural exporting countries (Argentina, Australia, Canada, Chile, Uruguay and the United States); the European Union; the Like-Minded Group (developing countries, excluding Argentina, Chile and

Uruguay); Central and Eastern Europe; and the Compromise Group, consisting of OECD countries that are not agricultural exporters nor part of the European Union (Japan, Mexico, New Zealand, Norway, Singapore, South Korea and Switzerland).

While the Miami Group was concerned about the impact of the protocol on the agricultural commodity trade, the European Union was responding to increased public concern about import into the Union of genetically modified foods. The Compromise Group with its eclectic membership represented a mix of such concerns, since it included leaders in biotechnology research such as Switzerland and major agricultural importing countries such as Japan. The central axis of conflict in protocol negotiations has been the push by LMO importing countries to have a broad scope for informed consent and broad national discretion in LMO decision-making.

While a broad scope has been pushed by developing countries, the European Union has sought to ensure discretion in decision-making. On the other hand, LMO exporting countries of the Miami Group have argued for a narrow scope for informed consent, and limited science-based decision criteria, to ensure that trade in LMOs is not unduly curtailed. These competing demands have been negotiated under the rubric of "informed consent" as the central governance mechanism of this regime.

The protocol's final obligations on the scope of "informed consent" (i.e. what should be included within it) represent a least common denominator agreement, given widely divergent demands in this area. The categories of LMOs debated for inclusion within informed consent were (a) LMOs for deliberate release into the environment; (b) LMOs for food, feed or processing (agricultural commodities); (c) LMOs for contained use; (d) LMOs in transit; (e) processed and finished products deriving from LMOs; and (f) LMO-pharmaceuticals. Developing countries argued initially for all LMOs and their products to be covered by informed consent.

Their rationale was that all LMOs posed risks to biological diversity and human health and thus merited information-sharing and consent prior to transfer. Most OECD countries argued that only LMOs for deliberate release should be covered by informed consent, since this was the only category of LMO likely to pose threat to biological diversity. Only LMOs for deliberate release are to be covered by informed consent, the category that all agreed should be included. Other categories of LMOs have information-sharing as the primary international obligation, and LMOs-pharmaceuticals are excluded from the protocol altogether, as long as they are being addressed in other international fora.

The obligation for "advance informed agreement" prior to transfer of LMOs for deliberate release is the center-piece of the Cartagena Protocol and the most far-reaching change from the status quo. The procedure is initiated by a notification of intended export of an LMO by an exporter to the competent authority of the receiving country. The notification has to include specific information about the LMO, as mandated under the protocol. The exporting country has to ensure that there is a legal requirement in place for accuracy of the information being provided.

Importing countries have ninety days to acknowledge receipt of the notification and 270 days to make a decision. These obligations have the concrete impact of shifting the burden of responsibility for initiating action to exporters of LMOs, from a status quo where such responsibility rested with an importing country. More explicitly, it shifts the burden of responsibility for action from those desiring information to those in possession of it. Thus, the obligation for informed consent prior to deliberate release of LMOs shifts the burden of responsibility for initial action from the user/regulator to the producer of an LMO.

The protocol also encourages capacity building to assist recipients of information to assess its relevance to their national contexts, and to make decisions based upon such assessments. In mandating a set procedure and time-frames for decisions, such as the 270 day limit, the protocol also attempts to harmonize and bring predictability to national decision-making about LMO transfers. The relevance of such obligations for national governance, to be explored in subsequent sections, will turn on how such shifts in burden of responsibility to share information and solicit consent relate to already existing biosafety rules and practices in particular national contexts such as India, and how attempts to standardize procedures and time-frames resonate with existing biosafety practices.

The protocol's obligations for LMOs for food, feed or processing (agricultural commodities) call for information sharing rather than solicitation of consent of an importing country. These obligations were contentious through the end, given the developing country demand to include such LMOs within the advance informed agreement obligation and Miami Group opposition to this demand. The compromise calls for LMO producer countries to notify the Biosafety Clearing House (the institutional mechanism under the secretariat to share information) of domestic approval of a new LMO within 15 days of the approval being granted.

A potential importing country can draw on this information to decide whether to restrict import of the particular LMO, which may enter international trade in the future. Countries can rely on their domestic biosafety regulations to make such decisions or use the protocol's procedure, whereby decision must be taken within 270 days of the domestic approval notification. These obligations serve, first, to validate a country's right to restrict import of LMO commodities under certain conditions. This is significant because, although countries have the sovereign right to take import restrictive decisions pursuant to their domestic regulations, such actions can be seen as inconsistent with obligations under World Trade Organization (WTO).

International validation through the Cartagena Protocol of the right to restrict commodity imports under certain conditions makes it less likely that domestic decisions will be seen as counter to WTO obligations. The protocol's obligations also place the onus on LMO producers to provide information about LMOs that could potentially enter international trade to the biosafety clearing house (which might otherwise be less easily accessible to different groups) A key difference, however, between the protocol's obligations for commodities and those for deliberate release is that the onus of responsibility to initiate action relating to transfer of commodities remains on importing countries

The responsibility rests with an importing country to ascertain which LMOs, from those reported to the Clearing House, might pose risk in their particular national context, and to assess the nature and extent of such risk prior to taking a decision about import. As with deliberate release obligations, there are provisions for capacity building and technical assistance to assist countries with this procedure. The relevance of these obligations for national governance requires, again, examining prevailing practices of biosafety information sharing in particular national contexts such as India.

The protocol's obligations for LMOs for contained use entail only information sharing about such LMOs to accompany all transfers. However, in a concession to varied understandings of what constitutes containment in different national contexts, the protocol allows that "standards in the Party of Import" will determine which LMOs are to be considered for "contained" use. This deference to national differences is one of the most striking illustrations of how global efforts to standardize safety practices are necessarily limited by the fact that "safety" remains a context-dependent concept (i.e. it will depend upon varying capacities to ensure containment, for example).

One outcome of the protocol's explicit deference to differing national understandings of containment is a reduced transnational predictability regarding the procedures governing transboundary movements of this category of LMOs. Given this (justifiable and inevitable) lack of standardization, the relevance of the protocol's obligation to share information about LMOs for contained use will turn on national understandings of containment, the ability to ensure adequate containment, and the already prevailing practices in biosafety information generation and sharing for such LMOs in different national contexts.

In addition to mandating a minimum scope for informed consent, the protocol's final obligations are also characterized by ambiguity regarding the nature and "sound-scientific" content of the biosafety information to be generated, and the decision-criteria for consent. Disputes in this area during protocol negotiations centered around whether consent was to be based upon "sound" scientific evidence of harm posed by LMO transfers or whether countries had the right to restrict such transfers in the absence of scientific certainty of harm (i.e. whether precautionary decision-making was permitted). While the Miami Group argued for decisions to be based upon sound science, the European Union insisted that precautionary decision-making was essential in this area, given pervading scientific uncertainty about risks.

This dispute was tied to the contentious issue of the relationship of the protocol's obligations to multilateral trade rules, such as those of the WTO Agreement on the Application of Sanitary and Phytosanitary Measures (SPS Agreement). This agreement calls for national sanitary and phytosanitary measures relating to human, animal and plant health to be scientifically sound and allows for precautionary decision-making only on a provisional basis. Developing countries supported the European Union on the need for precautionary decision-making for LMO transfers, yet this was clearly an issue over which the faultlines were deeper between OECD countries. For countries of the South, a main concern was inclusion of socioeconomic considerations in LMO decision-making.

Developing countries argued that risks posed by LMOs went beyond quantifiable harm assessed through scientific risk assessments, even those that could account for scientific uncertainties. Key concerns relevant to LMO decision-making included whether widespread use of LMOs would fuel new forms of dependencies on technologically advanced countries or multinational companies, or affect traditional livelihoods. Most OECD countries opposed inclusion of socioeconomic considerations in LMO decision-making. They argued that such considerations varied greatly between countries and their inclusion would preclude predictable and harmonized international biosafety rules.

The Cartagena Protocol's final compromises call for decisions about LMO transfers to be based upon a scientifically sound risk assessment. Also allowed, however, are precautionary restrictions on LMO transfers in the face of scientific uncertainty about adverse impacts posed by such transfers. Finally, socioeconomic impacts can be taken into account in decision-making, yet their scope is narrowly restricted.

Science-based risk assessment

In calling for science-based decision-making for LMO transfers, the protocol mandates that decisions be based upon a quantitative risk assessment. Such risk assessments are to be "carried out in a scientifically sound manner...taking into account recognized risk assessment techniques". Importantly, the protocol's obligations on risk assessment concedes that the data to be generated in such assessments cannot be internationally harmonized, given the diverse agroecological environments within which they will be undertaken. As acknowledged in the protocol's obligations, in different risk assessments "the required information may vary in nature and level of detail from case to case, depending on the living modified organism concerned, its intended use and the likely potential receiving environment".

Scientists from different countries sought, for example, to define the ecological unit represented by the "receiving environment" for release of LMOs. Those arguing for a broad understanding of this term noted that a receiving environment should include not only the particular field into which an LMO was sown but also its surrounding areas, since these could inadvertently become receiving environments. The term "potential" was added to "receiving environment" to address this concern. However, this was seen as too broad by others, who argued that "potential receiving environment" for an LMO could be the whole planet. These scientists proposed the qualifier "likely" to address their concern. As a result, the ecological zone for which the scientifically sound assessment of adverse impacts from LMO releases is to be undertaken is characterized in the protocol as the "likely potential receiving environment".

The ambiguity inherent in this formulation leaves it to the discretion of each national risk assessor to determine the scope of the receiving environment for an LMO. This openness to differential interpretations is perhaps unavoidable, yet the assumption underlying the call for scientifically sound decision-making is that such determinations can be made on the basis of universally understood or uniformly interpretable scientific criteria.

This assumption was also revealed as problematic in attempts to develop a shared scientific understanding of the concept of "novelty" which underpins definition of a living modified organism.

When divergent scientific understandings of novelty could not be reconciled, it was left unexplained, thereby allowing differential interpretations to persist. It is argued, from the above, that the call for scientifically sound decision-making in an arena where the science itself remains heavily contested will inevitably result in ambiguity and openness to differential interpretations, rather than standardized, uniformly interpretable criteria. Clearly, the implications of this for nationallevel governance, to be explored in subsequent sections, will turn on the nature of the decision-criteria currently being relied upon, as well as whose understandings of "sound science" currently prevail.

Precautionary decision-making

As noted, in debates over sound science versus precautionary decision-making, the protocol legitimizes both as necessary bases for decision-making about LMO transfers. Thus, in addition to requiring a scientifically sound risk assessment, the protocol allows for potentially import-restrictive decisions in the face of scientific uncertainty about adverse impacts posed by LMOs (precautionary decision-making). The protocol's understanding of precautionary decision-making is included within its decision-procedures governing LMOs for deliberate release and LMO commodities. The language on precaution in the body of the protocol has been hailed as the first operationalization of the precautionary principle in an environmental agreement.

However, examining the language reveals that rather than operationalizing "the" precautionary principle (there is no universally shared version) it represents a mix of existing formulations in other agreements. A comparison with other dominant renditions of precautionary decision-making, including Principle 15 of the Rio Declaration and Article 5.7 of the WTO Sanitary and Phytosanitary Measures Agreement, reveals that the protocol's language can be differently interpreted to both exceed and not exceed what is allowed for under existing agreements with relevance for LMO trade. In particular, one reading of the language on precaution in the text of the protocol is that (in keeping with Article 5.7 of SPS) it takes a scientific risk assessment as its starting point.

The language specifies that precautionary actions are justified when there is scientific uncertainty about "the extent" of an adverse impact posed by an LMO (but makes no reference to uncertainties relating to whether or not an adverse impact exists). This can be interpreted as requiring prior scientific documentation of the existence and nature of an adverse impact before precautionary action can be taken. In addition, however, the protocol also evokes Principle 15 in its preamble and objectives. The language of Principle 15 is again open to multiple interpretations. While it imposes a criterion of "cost-effectiveness" on precautionary actions on the one hand, it also states that "lack of full scientific certainty" should not preclude action to mitigate harm. Such references raise the specter of a zero-risk standard for some, since "full" scientific certainty about lack of adverse impact is seen as unattainable.

In sum, the outcome is openness to multiple interpretations of the criteria which can trigger legitimately precautionary actions under the protocol. The implications of ambiguity in this area remain important to assess in different national contexts, given the increased attention to the precautionary principle in the emerging global environmental governance architecture. While much debate on precautionary decision-making has occurred within

OECD countries, the relevance of precautionary environmental health and safety standards for developing countries has been little examined. Even fewer studies have examined how precautionary decision-making is operationalized within national-level decision-making in developing countries such as India.

Socioeconomic considerations

Finally, the protocol's finalized decision-criteria allow for limited consideration of adverse socioeconomic impacts in decisions about LMO transfers. Specifically, it allows countries to take into account "consistent with their international obligations, socioeconomic considerations arising from the impact of LMOs on the conservation and sustainable use of biological diversity...". This is a narrow formulation insofar as it links the socioeconomic impacts that can be taken into account to impacts on biodiversity. This excludes considerations such as export substitution potential of LMOs, the loss of traditional livelihoods, or increased dependence on multinational companies, socioeconomic concerns voiced by developing countries. In addition, the proviso that decisions based on socioeconomic considerations are to be consistent with a country's international obligations (under the WTO) ensures that this concession to developing country concerns does not enhance national discretion beyond what the WTO already allows for.

The question for transnational-national governance, evaluated in subsequent sections, is the relevance of such a narrow formulation for countries such as India, where socioeconomic considerations figure prominently in public debates about LMO use and are an important component of biosafety decision-making. As the above analysis of the protocol's obligations reveals, its scope and criteria for "informed consent" are characterized by least common denominator agreement and by ambiguity. In particular, the scope of informed consent (what it should cover) is characterized by minimalist agreement, since only LMOs for deliberate release are included. In turn, the criteria for consent, in particularly scientifically sound as well as precautionary decision-making, are characterized by ambiguity and openness to differential interpretations.

National Governance of Biosafety in India

Use of Genetically Modified Organisms in Indian Agriculture

Although little systematic research has been undertaken about public perceptions surrounding biotechnology use in India, it can be ascertained from media reports and interviews that public concern about GMO safety is not currently widespread. Instead, debate is concentrated amongst a few vocal critics and proponents of GMO use in agriculture. The critics include non-governmental organizations, prominent among them the Research Foundation for Science, Technology and Ecology (RFSTE), led by activist Vandana Shiva. Alliances between groups concerned with biosafety and intellectual property rights are likely when the two issues intersect, such as when patented seed is field tested to meet biosafety requirements.

Opponents of seed patenting in India include a non-government organization, the Gene Campaign, launched by geneticist Dr. Suman Sahai in 1992 to oppose the trade-related intellectual property rights proposals in the Dunkel Draft of the Uruguay Round multilateral trade negotiations (Sahai, undated). Her campaign against the patenting of seed has received the support of highly influential farmers organizations, some of which are also opposed to transgenic crops. On the other hand, proponents of the use of biotechnology in agriculture include many leading public sector agricultural scientists and government regulators, especially at the Department of Biotechnology, whose mandate is to facilitate (but also to regulate) biotechnological research in India.

Key private sector seed companies, with a history of involvement in developing hybrid seed, also emphasize the need for transgenic crops in Indian agriculture. Finally, the Supreme Court can be an influential player in the Indian biosafety debate, given the potential for public interest litigation to be brought against government regulators alleging non-compliance with national and international obligations to ensure biosafety. Within these different groups, views about risks and benefits posed by GMOs are similar to those espoused elsewhere, i.e. they cover the spectrum of ecological, human health, food safety, ethical and socioeconomic concerns.

However, the issues that have generated the most impassioned debate in India have less to do with ecological or food safety concerns and more to do with socioeconomic considerations arising from increased reliance on GMOs in agriculture. The socioeconomic concern voiced most often is that reliance on transgenic seeds will exacerbate small farmer dependence upon multinational companies and capital intensive agriculture. Groups who oppose use of biotechnology in agriculture often cast their arguments in overtly nationalist idioms, with slogans such as "Monsanto Quit India" and "bija satyagrah" (seed-related civil disobedience), evoking images of the anti-colonialist freedom struggle of the early 1900s.

Socioeconomic concern over increased foreign dependence is linked to the always complex issue of food security in countries such as India. Food security in developing countries is evoked by supporters of biotechnology as a central reason to embrace transgenic crops, given the need to increase agricultural productivity in the face of a declining resource base. This claim is dismissed as disingenuous by opponents, who point out that hunger is not necessarily related to insufficient food production. Notwithstanding rhetorical references to food security in the debate on GMOs, a concern with food security is clearly of immediate salience for a country where close to 70% of the population relies on agriculture for its livelihood, and a majority live below the poverty line.

Another focus of genetic transformations has been production of higher-value hybrids for crops such as mustard. According to the largely private sector developers of this modification, such transgenic crops respond to a market opportunity and also meet a priority need, given that India currently imports large quantities of oilseeds. In addition, research mainly in the public sector is also underway on nutritionally altered transgenic crops, such as, for example, an effort to enhance protein content in potato, as well as a wide range of other applications such as moisture tolerant and delayed ripening crops. While research on GMOs is expanding at a rapid pace in India, no transgenic crop has yet been approved for commercialization.

Biosafety Regulatory Framework

Genetically modified organisms are regulated in India under the purview of the 1986 Indian Environment (Protection) Act (henceforth, the EP Act). The broad objective of the EP Act is the protection and improvement of the environment. To meet this objective, the Act calls for regulation of "environmental pollutants" which are defined as "any solid, liquid or gaseous substance present in such concentration as may be, or tend to be, injurious to the environment". The broad definition of "environmental pollutant" was used by the Ministry of Environment and Forests in 1989 to issue rules to govern use of genetically engineered organisms under the EP Act.

The 1989 "Rules for the Manufacture, Use, Import, Export and Storage of Hazardous Microorganisms,

Genetically Engineered Organisms or Cells" (henceforth 1989 Rules) constitute the legally binding regulatory framework for GMOs in India. As evident from the title, GMOs are placed here in the same category as hazardous microorganisms and their regulation under the EP Act is justified by their alleged potential to be hazardous substances or environmental pollutants. Since the 1989 Rules call for guidelines to be developed to give them effect, biosafety guidelines were issued by the Department of Biotechnology under the Ministry of Science and Technology in 1990, and have been consistently revised and expanded in the last decade.

Development of this regulatory framework dating from the early 1990s has been in response to the need for regulatory oversight of the growing domestic community engaged in biotechnology research in both the agricultural and pharmaceutical sectors. It was also in response to debates within the European Union about safe use of genetically modified organisms, during negotiation of regional directives on contained use and deliberate release of GMOs. The possibility that India might be a future importer of GMOs from OECD countries thus also spurred development of a domestic regulatory framework.

The biosafety regulatory framework consists of the 1989 Rules issued by the Ministry of Environment and Forests, as elaborated and revised by the 1990, 1994 and 1998 Guidelines issued by the Department of Biotechnology. These cover the entire spectrum of activities relating to genetically modified organisms. This includes "research involving genetically modified organisms, as well as genetic transformations of green plants, rDNA technology in vaccine development, and large-scale production and deliberate/accidental release into the environment of organisms, plants, animals and products derived from rDNA technology".

Production facilities such as distilleries and tanneries which use genetically modified organisms are also covered. The 1990 "Recombinant DNA Safety Guidelines" and 1994 "Revised Guidelines for Safety in Biotechnology" provide detailed guidance on containment and safe laboratory practices for GMOs in the agricultural and pharmaceutical sectors. They also, however, contain an important change from the 1989 Rules in their treatment of deliberate release of GMOs. While the 1989 Rules effectively banned such releases (permitting them only under special circumstances), the 1990 Guidelines permit them, with a shift in focus to assessing and managing ecological and health risks that might result. This is similar to the evolution of GMO regulation in OECD countries such as the United States, where self-regulation by scientists in the early 1970s prohibited deliberate release and focused on ensuring safe use of GMOs in contained conditions, yet where the prohibition on release was revoked in a relatively short period of time.

The 1998 "Revised Guidelines for Research in Transgenic Plants and Guidelines for Toxicity and Allergenicity Evaluation of Transgenic Seeds, Plants and Plant Parts" apply only to transgenics for use in agriculture (in contrast to the 1990 and 1994 rules which covered both the agricultural and pharmaceutical sectors). The latest rules add to the regulatory architecture by calling for toxicity and allergenicity data on transgenic plants and plant parts. They also clarify how the procedures governing import of genetically modified organisms for research are distinct from imports for release or commercialization.

However, notwithstanding this comprehensive coverage, GMO use in India to date has occurred largely in contained conditions, with some deliberate releases in the form of experimental field trials. Approvals for commercialization of GMOs or food safety assessments of processed materials have not yet been required.

Furthermore, no transgenic products have knowingly been imported into the country for commercial use. Thus, while the biosafety regulatory framework in India is broad in its coverage, its functioning in practice remains in a state of evolution and development.

The information to be taken into account in assessing GMO safety is contained in the 1990, 1994 and 1998 Biosafety Guidelines. Information requirements are very similar to those in risk assessment models elsewhere. Indian risk assessment guidelines have drawn upon models used by the United States Department of Agriculture Plant and Animal Health Inspection Service (APHIS) as well as from biosafety guidelines elaborated by other OECD countries. Guidelines developed by international agencies, such as the Organisation of Economic Cooperation and Development (OECD) and the World Health Organization (WHO) have also served as models.

Information taken into account or required to be generated in a biosafety assessment includes details of the modified organism, such as description of the host; source of the transgene; characteristics of expression vectors, insertion genes and promoters; transformation methods; and a genetic analysis of stability and biochemistry of the expressed product. For laboratory use and greenhouse trials, information is required about backcrossing methods, germination rates, phenotypic characteristics, and toxicity and allergenicity potential in handling the modified organism. For field trials, information is required about germination rates, gene flow, invasiveness potential, possibility of weed formation, possibility of transfer of transgenes to near relatives, and toxicity and allergenicity potential. Also, data is to be generated on the long-term survivability of the novel organism, including susceptibility to diseases and pests, as well as a comparison between the modified and non-modified organism in pest susceptibility. Finally, data on comparative economic benefits of the modified plant is also required. Although Indian evaluation criteria are not distinct from risk assessment models elsewhere, it does diverge from such models in one significant manner.

As noted above, a recently added requirement in safety assessments includes an agronomic evaluation of the transgenic crop to determine economic advantage to farmers. This addition to the 1998 Biosafety Guidelines now constitutes an integral component of the technical risk assessment, along with the ecological and human health safety evaluation. It is required before a "biosafety clearance" can be issued to a transgenic crop. The generation of economic viability data during the technical safety evaluation is centrally linked to the socioeconomic dimension to biosafety in the Indian context. This inclusion of socioeconomic data in the biosafety calculus is reflected, as well, in the recent decision in July 2000 to grant permission for large-scale field testing of Bt cotton in India (the first crop to receive such approval). Although biosafety protocols for large-scale testing are still being drawn up, data required to be provided in this case includes the "cost of transgenic seed, projected demand, and the area to be covered under transgenic cotton cultivation".

In India, much debate has centered around the legitimacy of consent givers for GMO use in agriculture. As per the 1989 Biosafety Rules, the authority to regulate GMOs is divided between the Ministry of Science and Technology and the Ministry of Environment and Forests. Research on GMOs is to be overseen by the Review Committee on Genetic Manipulation (RCGM) under the Ministry of Science and Technology's Department of Biotechnology. Deliberate release and commercialization of GMOs is to be overseen by the Genetic Engineering Approval Committee (GEAC) under the Ministry of Environment and Forests. Although the functions of these two national biosafety committees now appear to be clearly delineated, the division of responsibility between them has been a source of much controversy.

Disputes have centered around where the boundary between research and deliberate release lies and, in particular, whether field trials constitute a research activity or a deliberate release of GMOs into the environment, since, if the former, they would be regulated by the Department of Biotechnology, and if the latter, by the Ministry of Environment. In addition to these national-level regulatory committees, every institution engaged in genetic engineering research in India is required to establish an Institutional Biosafety Committee. Furthermore, State Biotechnology Coordination Committees and District-Level Committees are to be set up to facilitate information exchange between the center and the states.

The most recent addition to this institutional framework is a Monitoring and Evaluation Committee to oversee the agronomic evaluation of the transgenic crop during field tests and to monitor biosafety data generation. Finally, a Recombinant DNA Advisory Committee is to meet occasionally to review national and international developments in biotechnology and recommend appropriate biosafety regulations for India.

Scientific disciplines represented include genetics, molecular biology and the agricultural sciences. There are, however, almost no social scientists and no members of the general public involved. Representatives from industry and non-governmental organizations can be invited to participate in their individual capacities as experts, although there is no formal requirement to involve them, and petitions by NGOs to participate in particular sessions have in a few cases been turned down. Also, these two central committees, but particularly the Department of Biotechnology's RCGM, are composed of public sector scientists who are themselves engaged in transgenic research. This results in the not unfamiliar situation of scientists regulating themselves rather than an autonomous agency regulating research in this area.

Equally important, public sector scientists are also regulating their private sector counterparts, who are often further along in developing similar transgenic crops. This public-private information sharing dynamic has implications for reliance on information sharing as a central risk mitigation strategy in this area. The role of other relevant ministries in biosafety regulations, such as those of agriculture and health, remains ill-defined. It is unclear, for example, whether transgenic seed is to be governed under biosafety rules alone or whether and how the 1966 Indian Seed Act also applies.

The ministry of agriculture is considering amendments to Indian seed legislation to cover transgenic seed. A particular concern is ensuring seed purity, i.e. ensuring that use of transgenic seed, if and when available, does not contaminate regular seed lines. This is an issue which the ministry of agriculture clearly sees as within its regulatory domain and outside the competency of the department of biotechnology or the ministry of environment. Another issue still to be resolved, yet with crucial implications for biosafety, is whether transgenic seed once de-regulated is to be treated as regular seed or whether it will require distinct seed varietal registration procedures from those in place for non-transgenic seed.

Current varietal registration rules in India offer two routes for placing new seed on the market: testing of seed and certification of efficacy through "all-India coordinated trials" administered by the public sector agricultural research system, or the alternative option of "truthful" labeling of new seed to be placed on the market. The debate at the moment turns on whether the "truthful labeling" option, historically preferred by the private sector for speedy entry into the market, should be permitted for transgenic seeds or whether the all-India coordinated trials should be made mandatory. There is also little clarity to date regarding jurisdictional authority for human health and food safety concerns raised by GMO use in agriculture.

The 1954 Prevention of Food Adulteration Act does not specifically cover transgenic entities. However, this is dependent upon how broadly food adulteration is understood and whether transgenic food additives can be considered adulteration. Again, the Ministry of Health is engaged in a process of internal consultation to determine its role in regulating transgenic food once available.

While the health ministry is the lead ministry responsible for negotiating labeling requirements for genetically modified foods within the Codex Alimentarius Commission (a United Nations standard setting body jointly established by the Food and Agricultural Organization and the World Health organization), they have not been involved with regulation of GMOs under the Cartagena Protocol. At the domestic level, however, representatives from the ministries of agriculture and health are included in national biosafety regulatory committees. Thus, their concerns can, in principle, be accommodated during deliberations of these committees in regulating domestic use of GMOs in India.

Controversies over use of Genetically Modified Organisms

A number of controversies in recent years regarding (a) approval to field test transgenic crops; (b) whether terminator technology was being imported into India; and (c) approval to import agricultural commodities containing GMOs; have clarified how existing rules work and have contributed to their evolution. The nature and role of these controversies in shaping the biosafety governance regime in India is described below.

Regulatory approval for field testing transgenic cotton in India

The question of how well the regulatory system works in

India received sustained scrutiny in the wake of a public interest litigation filed by the Research Foundation for Science, Technology and Ecology (RFSTE) in the Supreme Court of India against the Department of Biotechnology, the Ministry of Environment and Forests, the Ministry of Agriculture, Mahyco, i.e. Maharashtra Hybrid Seeds Company (an Indian private sector seed company), and Mahyco-Monsanto Biotech India Ltd (a joint venture established between Monsanto and Mahyco). The RFSTE alleged that improper authorization was given to field test Monsanto/Mayhco's transgenic Bt cotton in India, and that existing biosafety regulations did not adequately protect against adverse ecological and human health effects posed by such transgenic crops.

While the "facts" associated with this field testing remain disputed, it can be ascertained from media reports that Mahyco acquired a Bt toxin gene from Monsanto in 1995, backcrossed it into Indian cotton crop varieties, and requested approval to field test its transgenic Bt cotton seed. During this same period, Monsanto acquired a 26% stake in Mahyco. Permission to conduct 40 field tests in 9 states was granted by the Department of Biotechnology's biosafety committee to Mahyco in 1998 but the tests have remained mired in controversy. In the public interest litigation brought by the RSFTE, it is alleged that these field tests constituted a deliberate release into the environment, and hence consent should have come from the ministry of environment rather than from the department of biotechnology.

However, those soliciting and giving approval, namely the private sector developers of the crop and governmental regulators, viewed the field tests as "experimental research in contained conditions" rather than as deliberate release. As a result of this controversy, an amendment to the Biosafety Guidelines now clarifies that the Department of Biotechnology's committee, the RCGM, has the authority to approve "small experimental field trials" limited to a total area of 20 acres in multilocations in one crop season, with any one location not exceeding one acre. Field trials exceeding these limits are to be considered large scale releases and would require approval from the ministry of environment. Notwithstanding this clarity about who has jurisdictional authority over which trials, it is revealing that the department of biotechnology's authority to consent to field tests was challenged in the Supreme Court when it consented to field tests of what was seen as Monsanto's Bt cotton.

Earlier approvals by the RCGM to field test domestically produced transgenic crops have not been challenged. The controversy over the appropriate consent-giving authority appears linked then to broader concerns of some NGOs over multinational control of transgenic crops in Indian agriculture, rather than to concrete evidence that the ministry of environment's biosafety calculus would differ from that of the department of biotechnology, which remains to be put to the test. The public interest litigation also alleges that only central government approval was given for the fieldtesting, without consultation with regional and state governments and local communities.

In some states, no State Biotechnology Coordination Committee mandated by the 1989 Biosafety Rules had been set up, and state and district-level authorities were unaware that transgenic cotton was being field tested within their territories. A state biotechnology committee was only established in the southern Indian state of Karnataka after the tests were well underway. Karnataka and neighboring Andhra Pradesh have been the site of protests by farmers against transgenic crops, spearheaded by the Karnataka farmers' union headed by controversial anti-globalization activist M.D. Nanjundaswamy.

Setting up a state biotechnology coordination committee in 1999 was portrayed by the Karnataka state government as a major step in enhancing vigilance over transgenic crops, even though such a committee was to have been established in the early 1990s according to biosafety regulations. More generally, this highlighted that despite the laws on the books, in many cases the state and districtlevel committees called for are not yet in existence. Moreover, where they do exist, they do not necessarily include scientists conversant with the technology under consideration or its use in agriculture.

From the perspective of private sector developers of transgenic crops, requesting state government approval has proved to be one of the most frustrating components of meeting biosafety regulatory requirements, also because there are no fixed times for state-level biosafety committees to meet. They meet on an as-needed basis, placing the onus on each GMO producer to convince the committee of the need for an expedited evaluation of their proposal for research or field-testing. In response to such concerns, biosafety regulators defend the existence and intent of the center-state participatory approval structure, and note that the system will adjust and evolve as the first transgenic crops move through the regulatory pipeline.

In addition to consent-givers, the information to be generated and procedures for field testing the transgenic crop were also disputed in the RSFTE case. The public interest litigation alleges, for example, that a comparative study of pest incidence in transgenic and non-transgenic fields is mandated by biosafety regulations, yet no such data was generated. It also alleges that mandatory containment measures as outlined in the biosafety guidelines were not followed. Thus, it alleges that although isolation distances of 5 meters around plantings of transgenic material are required, such distances were not maintained. These allegations are disputed by the Department of Biotechnology, yet they highlight potential constraints to field-testing transgenic crops likely to be faced in the Indian context.

Thus, maintaining and monitoring 5 meter isolation distances in very small farmer holdings may not always be feasible, notwithstanding whether it is a technically adequate containment parameter. Although the public interest litigation remains pending, it has already had a discernible influence on the biosafety governance regime in India. In addition to generating public debate, it has also clarified the functioning of the consent giving process for GMO field trials.

Terminator technology

Together with debates over field-testing transgenic cotton, a period of sustained controversy in late 1998 and 1999 centered around the alleged testing of "terminator technology" in India. Called "gene protection systems" by its developers, this technology entails modification to a crop such that it produces sterile seed. The objective is to prevent farmers from saving transgenic seed for use the following year. This is defended by proponents of the technology as a necessary biological method of intellectual property protection and is attacked by opponents as depriving farmers of an age-old right to save, share and exchange seed.

The origin of the allegation that terminator technology was being tested in India is unclear, yet it became tied to protests surrounding field testing the transgenic Bt cotton. This seems to be the result of a similarity in timing between the field tests and Monsanto's move in the United States to acquire the seed company Delta Pine Land and Co. which holds a patent on a type of terminator technology.

Relevance of Cartagena Protocol

In examining the implications of the Cartagena Protocol's rules and obligations for biosafety governance in India, a preliminary hypothesis emerging from the analysis of the

Cartagena Protocol can be postulated. The rules and obligations of the Cartagena Protocol will not change the prevailing status quo for biosafety governance in India. Thus, with its minimalist scope, it is not likely to exceed the categories of LMOs requiring consent in Indian regulations; and with its ambiguous decision criteria, it is likely to leave national discretion in LMO decision-making unchanged. This is at first glance a relatively uninspiring hypothesis, in light of the long drawn out and contentious nature of these multilateral negotiations.

If the completed protocol leaves unchanged the scope and decision-criteria of domestic biosafety regulations, should such a regime be considered irrelevant or ineffective? Clearly, relevance and effectiveness have to be evaluated in light of the objective of a regime, especially from the perspective of a developing country and LMO importer such as India. If an original intent was to negotiate a regime which would defer to national differences in LMO decision-making, then in leaving domestic biosafety regulations unchanged (and in not limiting them) the Cartagena Protocol accomplishes this original intent. This can be construed as a measure of its "effectiveness" and by this measure, it succeeds.

At the very least, the Cartagena Protocol embodies within it a norm that LMOs require special regulatory attention. This was contested through much of the early stages of the negotiation, given divergent views on the uniqueness of risk posed by LMOs. Furthermore, the protocol also legitimizes the restriction of trade in LMOs under certain conditions. Until the process of negotiating a protocol commenced in the mid-1990s, there was no explicit international norm on the need to regulate such trade. Thus, attempts in this period to restrict LMO transfers pursuant to domestic regulations could be seen as hostile to a de facto prevailing international norm of the unhindered transfer of such entities.

The relevance of a completed protocol for the transnational-national biosafety interface is then that it serves first and foremost to legitimize the existence of domestic biosafety regulations. Clearly, however, while the protocol lends legitimacy to domestic regulation of LMOs, can it simultaneously legitimize contradictory approaches to LMO regulation, embodied within different domestic frameworks? However, in an increasing number of international negotiations, including biosafety, the source of conflict centers less around devising new obligations and more on whose already existing national approaches will be internationalized. How such conflicts are resolved will be crucial to the relevance of the final agreement for distinct national contexts.

Although internationalizing their domestic approaches was most characteristic of the intra-OECD conflict in protocol negotiations, it was also an important concern for India. Thus, Indian negotiators sought to ensure that a completed protocol would not limit the already existing scope or decision-criteria for biosafety governance in domestic regulations.

Private and Public Interface in Biosafety Information

In the case of India the domestic private sector is at the forefront of developing and testing transgenic crops and in generating biosafety data. It is also sharing such data with regulatory authorities composed of public sector scientists, many of whom are also engaged in transgenic research. This private-public dynamic in information sharing brings to the fore the important but less examined link between intellectual property rights (IPR) concerns and biosafety. Given that privately held information about transgenic technologies is revealed during the biosafety assessment process, two sets of concern arise: a concern with legitimacy for those on the receiving end of biosafety information, and a concern with ensuring confidentiality for those on the giving end.

For private sector generators of biosafety data in the Indian context, the confidentiality concerns can relate to the perceived lack of adequate domestic intellectual property laws, especially given that information is being shared with potential public sector competitors. This latter concern could be partly ameliorated if regulatory authorities in India were autonomous watchdog agencies instead of public sector scientists engaged in similar research. Confidentiality concerns can influence both the transgenic technology chosen to be deployed by the private sector and the biosafety assessment data that is shared.

The impact on choice of technology is reflected in the fact that the private sector has chosen to focus largely, in countries such as India, on development of transgenic hybrids rather than open-pollinated crops, since intellectual property concerns are less salient for hybrids. Even so, all centrally important components and production processes in a transgenic crop (whether hybrids or open-pollinated varieties) are likely to be patented or considered "confidential business information". As noted by the general manager of ProAgro PGS (a private sector company at the forefront of developing transgenic crops in India), in any transgenic crop, the plant variety germplasm, the selectable marker gene, the novel gene's trait, promoter and coding sequence, the transformation technology and the gene expression technology are all patentable, where permitted under intellectual property regimes.

In such circumstances, especially if patents have not yet been acquired, confidentiality concerns can impact the information that is willingly shared by private sector developers of transgenic crops. At a minimum, concerns over confidentiality can affect the information available to a broader public, even if it has to still be provided to biosafety regulators. Under such circumstances, the onus is even more strongly upon such regulators to ensure an accountable and participatory national-level biosafety decision-making process, a requirement that the Cartagena Protocol's deference to national "competent authorities" cannot ensure.

The protocol briefly addresses the relationship between intellectual property concerns and biosafety in an article on "confidential information" which remained contested during its negotiation. While developing countries called for its complete deletion, the finalized article responds to GMO exporting countries' concern that information shared under the protocol should be kept confidential if so desired by its providers. Thus, it calls on importing countries to ensure that there are procedures in place to protect confidential information. In response, however, to the concern that confidentiality requirements can stymie an adequate biosafety assessment, the protocol also mandates that certain data, such as a "general description of the living modified organism" and "a summary of the risk assessment" cannot be considered confidential.

This can be an important advance in contexts where even such limited obligations to share information do not exist. Especially in conjunction with the additional requirement that an exporting party must "ensure that there is a legal requirement for accuracy of information provided by the exporter" the Cartagena Protocol has the potential then to serve as an "international freedom of information act" or an international "right to know" law. This can be an important function in a realm where the legitimacy of information often depends upon its source.

In India, biosafety data provided by companies such as Monsanto is often viewed as suspect by certain groups. If so, the protocol's stipulation that certain information cannot be considered confidential, as well as the obligation to ensure its accuracy, can serve as a reassurance mechanism. What remains uncertain, however, is whether such shared biosafety information adequately reflect contestations over risk or contains the information necessary to undertake adequate risk evaluations elsewhere. A striking illustration of the potential hurdles to risk evaluation posed by the interface between biosafety and IPR concerns is contained in a recent media report on genetically modified trees.

The protocol's relevance for biosafety governance will clearly turn during its implementation phase on whether it provides for a more open sharing of information than is the case to date. In countries such as India with evolving biosafety rules, another broad implication of the private-public interface in information sharing is that standard-setting for biosafety is being undertaken through close interaction between private sector providers of information and public sector regulators. Many amendments to Indian biosafety regulations have been in direct response to evaluating and authorizing the first field tests of transgenic crops, many of which have been privately developed.

Private sector developers of such crops have therefore interacted closely with biosafety regulators in determining adequate safety parameters for field-trials. Thus, in India, seed companies such as Mahyco or ProAgro PGS have been involved not only with soliciting consent but also with development of standards for biosafety assessments. This can again raise concerns about the legitimacy or stringency of the resultant biosafety standards, and would certainly do so if the transnational private sector alone were playing this role. Concerns over involvement of the private sector in biosafety standard setting appear to be less pronounced in such cases.

If credibility concerns can be addressed, involvement of the private sector in biosafety standard-setting can, in the Indian context, also be perceived as a source of learning for public sector biotechnological research. This is especially the case in light of the declining resource base for public sector agricultural research. Since the domestic private sector is privy to the latest technological innovations yet is being regulated by public sector scientists, the learning resulting from field testing and data generation for transgenic crops is accessible to the public sector research establishment as well. Such learning can set the stage for public-private partnerships in the area of transgenic research, often advocated by public sector scientists and policy makers as a way to offset intellectual property hurdles and resource inadequacies.

In particular, the need for such partnerships can turn on the comparative advantage of the private sector in converting basic research to commercializable products desired by farmers. The public sector has tended to lag behind the private sector in this area, a situation that is likely to be exacerbated in the case of transgenic research. This is because, although the strength of the public sector lies in its ability to conduct basic research, the infrastructure required to commercialize such research, and in particular, to undertake biosafety assessments prior to commercialization, is currently either lacking or uncoordinated across public sector institutes. Thus, public sector transgenic research in isolation runs the risk of moving from one "basic research" project to another, with no long term perspective on how such projects fit into the larger agricultural research goal of ensuring food security for the neediest sections of the population.

8

Environmentally Responsible Use of Biotechnology

Biotechnology is one of the tools that will help solve our shared and increasingly critical problem of ensuring an adequate and sustainable food supply for all people. It opens up new possibilities for improving crops and livestock for higher yield and quality, enhanced nutritional content, improved storage and processing characteristics, and many other desirable traits. It also provides a means for developing alternative uses for agricultural crops, and even remedying environmental problems through bioremediation. Agricultural research managers charged with implementing biosafety policy play a major role in ensuring the environmentally responsible use of this powerful technology.

Biosafety is associated with the use of genetically modified organisms (GMOs) and, more generally, with the introduction of nonindigenous species into natural or managed ecosystems. A relatively new concept in agricultural research, it tempers the adoption of a new technology by carefully considering its potential effects on human health and the environment. Biosafety is achieved by assessing and managing environmental risks, evaluating potential ecological consequences, and weighing these against potential benefits. A step-by-step, case-by-case approach fosters a deliberate, informed, and environmentally responsible use of biotechnology. Policymakers begin by naming a lead agency under which biosafety policy is to be implemented. The next step is usually to draft biosafety guidelines that provide a framework for administrative procedures and decision making regarding the appropriate use of GMOs.

Managers charged with implementing biosafety policy can do so more effectively if they think in terms of instituting a biosafety system, rather than simply writing a document. Even though the guidelines are the most visible part of all biosafety systems, three other components are equally important:

(1) the people involved,

(2) the review process they implement, and

(3) mechanisms for feedback.

All well-functioning biosafety systems have these four elements in common, though they may take different forms and have different degrees of importance in implementation. Taking this broader view, a system must be put in place in which the four elements function together to produce environmentally responsible decisions. Each element presents its own management challenges:

— The guidelines formulated should be transparent, science-based, and flexible.

— Competence and confidence must be built in the people who are involved.

— The biosafety review process must be appropriate.

— There should be means of ensuring feedback so that the system improves through experience.

These elements and the challenges that they present are related and interdependent. People who conduct biosafety reviews do so within the regulatory framework articulated in the written guidelines. The extent to which the guidelines are science based and clear is reflected in the quality of decisions that they take. These decisions, in turn, are a product of the competence and confidence of reviewers and applicants. All are supported by accumulated experience and new information that is fed back into the system. Efforts to strengthen any one part ultimately strengthen the entire system and all the biosafety decisions coming from it. In the same way, neglect of any one part will weaken the whole system, hampering sound decision making.

Elements of a Biosafety System

Guidelines

Biosafety guidelines set forth policies and procedures for ensuring the safe use of biotechnology and its products. Ideally, the document clearly defines the objectives and scope of regulatory authority, specifies the allocation of responsibility within government agencies, lists the duties and membership of national and institutional biosafety review committees, and describes application and review procedures. Guidelines should stipulate that scientific principles are the basis for risk assessment. They should be flexible so that recommendations for risk management are commensurate with the level of identified risk.

The inherent flexibility of biosafety guidelines determines the extent to which the review process can, over time, be increasingly focused on cases that may pose some risk. As experience and familiarity increase, the same flexibility would allow a gradual relaxation of regulatory oversight for those releases that are found to pose little or no risk to human health or the environment. To draft guidelines that meet these standards, managers must consider a number of factors, including the statutory form, scope of oversight, objectives, and the stakeholders.

Statutory Form

Biosafety policy can be implemented in the form of new legislation, adaptation or extrapolation of existing legislation, or through nonlegislative means such as a ministerial decree. In deciding which form to use, managers must weigh the merits and drawbacks of each, for example, in complexity, timeliness, flexibility, and cost. While a new law carries the weight and enforcement power of government regulatory authorities, it is difficult to amend, and a lengthy and expensive effort is needed to enact and implement it.

A ministerial decree, in contrast, is faster and simpler to issue, and it is more readily amended or replaced. Without regulatory authority to enforce compliance, however, it may lack substance. Adaptation of existing laws, as was done in the USA, avoids the necessity of drafting new laws but can lead to redundancy and delays, for example, where a product such as Bt potatoes cannot be marketed until cleared by three separate regulatory agencies.

A biosafety system needs to be compatible with other regulatory authorities and procedures. For example, importation of seed, commodities, embryos, livestock, or biocontrol agents may be subject to plant protection and quarantine rules or regulations on animal health or food safety. Administrative jurisdiction, application procedures, and record-keeping requirements pertaining to biosafety should be coordinated with existing authorities in order to minimize duplication and conflicts.

Some countries, Australia being a good example, have developed guidelines in a form that serves a practical purpose in helping researchers and reviewers in risk assessment and risk management procedures. Applicants requesting approval for a GMO release are prompted to answer comprehensive and detailed questions about the proposed release, thereby providing the information necessary for evaluating potential risk.

Scope

Guidelines should clearly articulate what is subject to biosafety review and what is not. Does the system apply to products of modern molecular techniques such as chromosomal manipulations and embryo rescue? Or does it apply only to research involving recombinant DNA methods? Are all agricultural plants, animals, and microorganisms considered? Are laboratory and greenhouse experiments included, or only releases into the environment?

Will the guidelines address vaccines, pharmaceuticals, or industrial feedstocks such as lubricating oils produced in recombinant animals or plants? What about nonagricultural applications such as industrial fermentation technologies? Will there be provisions for evaluating commercial-scale releases? Will the safety of genetically engineered foods be evaluated the same way as nonbiotech foods? Answering these preliminary questions enables managers to determine the size of the task and estimate the resources they need to complete it.

Objectives

A careful analysis of objectives can be one of the most useful preliminary steps in designing a biosafety system. By definition, the primary objective of biosafety systems is to ensure the environmentally responsible use of biotechnology in order to reap its benefits without causing significant, unintended negative effects on people or the environment. There may be secondary objectives, however, which are more immediate. Having a policy and procedures in place may be seen as a way for developing countries to secure access to biotechnology products.

When a multinational company is seeking to field-test GMOs, the initiative to develop guidelines may stem from a pragmatic decision to quickly get something on the booksî so that an opportunity for technology transfer will not be lost. Instituting a biosafety system may also serve political objectives. Examples are asserting national autonomy, facilitating international trade, and participating in global initiatives such as the Convention on Biological Diversity. The recent failure of the open-ended ad hoc working group on biosafety under the Convention on Biological Diversity to reach agreement on an internationally binding biosafety protocol puts added pressure on individual countries to develop their own guidelines.

Stakeholders

Biosafety policies and procedures affect a diverse group of stakeholders. Scientists and project leaders engaged in field research are directly subject to biosafety regulations. The private sector also has a stake in how the guidelines address the use of GMOs. Local as well as multinational companies want a regulatory environment that is transparent, reasonable, and consistent. They have a stake in how commercial use of GMOs will be regulated in terms of application and approval procedures, environmental safeguards, and consumer issues such as labeling, in addition to matters of trade and importation. For biotechnology regulators, biosafety is typically an added responsibility that must be integrated into their other duties.

Government agencies concerned with biotechnology applications, typically including ministries of environment, health, and science and technology, and regional or local authorities have vested interests in how the system is established and run. The public's concerns about biotechnology's environmental effects and food safety are based in part on past experience with scientific advances whose immediate benefits were trumpeted before longer-term negative effects were realized. This and concerns raised by environmental groups and consumers' organizations need to be addressed. One way to secure public input is to have a respected member of the community (e.g., a local official, medical specialist, teacher, or extensionist) serve on the local institutional biosafety committee. A system that actively seeks public input and openly addresses the risk-benefit issues is, in the long run, one of the most effective ways to build public acceptance of biotechnology and its products.

People

The people working in a biosafety system are the public- and private-sector scientists seeking to test GMOs and members of the review committees who decide on proposed releases. The applicant and members of the biosafety review committee seem to have somewhat different roles, but they are engaged in a cooperative effort with a shared goal:

— The research scientist prepares a proposal to test transgenic plants in the field.

— Members of the institutional biosafety committee review the researcher's application.

— Members of the national biosafety committee are responsible for granting approval.

All those involved need to be familiar with the environmental issues associated with biotechnology products and have a working knowledge of the biosafety review process. They need to recognize what constitutes a potential risk and what risk management strategies may be applicable. Because advances come quickly and biosafety concepts are in constant evolution, all these people need ready access to the latest scientific, regulatory, and biosafety information.

Review Process

A biosafety review systematically evaluates a GMO, the site where it will be released, and the conditions under which the release will be conducted. The process is governed by the idea that, just as with any technology or human endeavor, there is no such thing as zero risk. Once the potential risk is identified, appropriate management procedures are specified to reduce it to an acceptable level. The emphasis of most biosafety reviews is on potential risk. However, applicants and biosafety review committee members also need to consider the potential benefits of allowing use of the GMO. Here they must recognize that not using a particular GMO may imply risks as well.

The review process begins with a proposal to introduce a GMO into the environment. (In some countries, greenhouse experiments with GMOs are also subject to biosafety committee review and approval.) The applicant has primary responsibility for assessing possible impacts or environmental consequences of a proposed release by comparing the GMO with the nonmodified or parental organism. This evaluation takes into account

- the nature of the organism that receives the new trait
- the donor organism from which the trait was derived
- the vector or mechanism used to transfer the trait
- the nature of the introduced trait, including potential toxicity of a gene product
- characteristics of the site or environment into which the GMO will be introduced
- elements of the release plan that provide containment or control

If a risk of sufficient probability and having consequences of sufficient magnitude be identified, the applicant determines suitable risk management measures that reduce risk to an acceptable level. This evaluation, conducted by the person most familiar with the GMO and documented in the release proposal, constitutes the first and most important level of biosafety assessment. An institutional biosafety committee conducts the second level of review. Generally, it evaluates the proposal for scientific content and verifies that the proposed release conforms to institutional and national guidelines. Institutional biosafety committee members collectively should have sufficient expertise in the relevant scientific disciplines to conduct an independent risk assessment and make appropriate risk management recommendations, if needed.

Final review and approval by a national biosafety committee may encompass nonscientific issues, such as accordance with quarantine regulations, international trade and treaty considerations, and socioeconomic impact. Not surprisingly, there are ongoing discussions over what constitutes an environmental risk and what is the limit for acceptable risk. Currently, biosafety reviews generally focus on a limited number of environmental issues associated with the release of transgenic crops. Two of these that concern the possibility that crops or their relatives may invade new territory, displace existing plant communities, or reduce species biodiversity. They may have added importance in regions that are centers of origin or diversity for the crop.

Weediness is potential for a crop to become established and to persist and spread into new habitats as a result of newly introduced genes is an issue when there is scientific evidence that acquisition of the new genes is sufficient to convert a domesticated species into a successful weed. Geneflow in which new genes are spread by normal outcrossing to wild or weedy relatives of the engineered crop becomes an issue if the new trait(s) confers a fitness advantage and becomes stably introgressed into the recipient genome. Toxicity is an issue associated with human health concerns over allergenicity and the safety of biotechnology foods and potential negative effects on nontarget organisms, especially beneficial species.

Pest and pathogen effects include a range of possible consequences such as the generation of novel viruses by molecular exchange within a transgenic plant, or emergence of target pest populations resistant to an engineered control mechanism. For many biotechnology products, none of these concerns is relevant. For example, based on long experience with conventionally bred pasteî tomatoes grown for processing, there is no evidence that tomatoes that are genetically engineered to have a higher solids content are more likely to become weeds than regular tomatoes. Also, there is no ecological evidence that increased solids confer a fitness advantage, and there is no reason to assume that pests or pathogens will be altered by their normal interactions with these GMO tomatoes. Issues of toxicity or allergenicity are not raised by increased amounts of normal fruit constituents.

If, however, a new marker gene or other gene product is involved, toxicity can be readily tested prior to release. It is incumbent upon applicants and reviewers to recognize those cases where a particular combination of crop, gene, and environment may present legitimate questions that need to be addressed before the GMO is grown in certain areas or on a large scale. Many decision support toolsî have been developed to guide reviewers through the biosafety review process. These charts, tables, decision trees, flow diagrams, etc. are not intended to supply yes-or-no answers, rather they can be used as roadmaps or checklists to ensure that all issues have been considered. An ongoing ISNAR research activity is also geared to help in this regard.

Feedback Mechanisms

New information and accumulated experience are essential inputs that allow biosafety systems to evolve. Feedback mechanisms are the built-in supply lines through which guidelines, people, and the review process keep up with the rapid pace of scientific advance and an accelerating rate of small- and large-scale releases of GMOs worldwide. Access to technical information and data gathered from prior releases is essential to support and strengthen subsequent biosafety decisions. At the same time, the accumulated experience of biosafety reviewers constitutes another source of feedback that can be used to improve oversight procedures. In a well-designed system, scientific and procedural feedback operate simultaneously to improve the quality of biosafety decisions.

Scientific feedback

Scientific feedback comes from external as well as internal sources. In evaluating a proposed release, biosafety review committee members may benefit from the experience of other countries by considering the acceptability of data from similar releases conducted elsewhere. Databases of field test information and biosafety reviews from the USA, Europe, and Australia can be useful to developing nations, so long as potentially significant differences in the environment, affected ecosystems, and agronomic practices are recognized.

Where there are specific gaps in knowledge concerning potential risks, issue-oriented meetings can be convened to clarify what is known and what management practices are appropriate. For example, there is some uncertainty about the potential to generate new plant viruses in transgenic plants expressing viral genes that confer virus resistance. Biotechnology regulators at the Animal and Plant Health Inspection Service of the United Stated Department of Agriculture (USDA) invited more than 20 prominent virologists to discuss the issue and propose experiments that would provide data to support more accurate risk assessment.

Review committees anticipating a need for data may specify reporting and record-keeping requirements as a condition of approval. Where appropriate, monitoring procedures specified as a condition of approval can be tailored to generate useful information. This type of feedback allows institutional and national biosafety review committees to gather information and identify emerging issues of local or regional importance. The relevance and usefulness of such feedback, however, will depend on the quality of data submitted. If, for example, monitoring to determine survival of a GMO or to detect transgene flow via pollen movement is required, it should be conducted in order to provide scientifically valid data. Subjective observations or anecdotal evidence are of limited value, but appropriate monitoring and reporting data can support subsequent decisions on proposed releases.

Procedural feedback

A periodic re-evaluation of guidelines and implementation procedures gives applicants, reviewers, regulators, and the public an opportunity to assess how well the system is working. Are applications reviewed on a timely basis? Is the burden of paperwork excessive or redundant? Can the review process be streamlined without compromising environmental responsibility? Are there new factors to consider, such as an administrative reorganization or a change in the budget? Is the public satisfied that their health and safety concerns are being properly addressed? Implementing biosafety policy does not end once guidelines are written, people are trained, and reviews are conducted.

It is a dynamic process that evolves through mechanisms for incorporating new information. As reviewers' experience and familiarity with introductions of GMOs increase, flexible regulations allow a gradual relaxation of oversight for those releases that pose little or no risk to human health or the environment. Limited biosafety review resources can then be focused on cases that do present some element of risk.

Human Resource Development (HRD)

Establishing a biosafety system imposes substantial HRD costs on the designated organizations or institutions. The major HRD cost is for various kinds of biosafety training. Loss of productivity can impose a substantial cost when researchers and managers are diverted from their regular jobs to draft guidelines and conduct biosafety reviews.

Site visits, facility inspections, and monitoring activities usually fall to regulators or institutional biosafety committee members who still have all their regular duties. As the number of proposals increases, additional personnel may be needed to serve as liaison to potential applicants, to handle applications and decision documents, to process reports, and to perform other information management tasks. Last, and perhaps most important to ensure long-term access to the benefits of biotechnology products, designated people will be needed to engage and inform the public.

Training

Biosafety training builds the competence and confidence of scientists, biosafety reviewers, and regulators alike. It enhances their awareness of environmental issues and potential consequences, and it provides a systematic approach to the evaluation of proposed GMO releases. A number of international agencies and programs support various forms of introductory and technical biosafety training at the institutional, national or regional level. Among these are the United Nations Industrial Development Organisation (UNIDO), the United States Agency for International Development (USAID), the Rockefeller Foundation, and the United Nations Environment Programme (UNEP).

Many government agencies are also active in biosafety education and training. Short and long training programs should be accessed on a regular basis. Biosafety, like biotechnology, is dynamic; new products, new technologies, and new concerns will continue to appear, presenting a moving target for regulators as well as research managers. There are one- or two-day programs that introduce current and future agricultural biotechnology applications, their potential benefits and associated risks, and how the biosafety system operates to guard against harm. These are useful to raise awareness and broaden the perspectives of senior people who influence policy decisions, manage research, or communicate with the public.

Longer, more intensive technical training gives members of institutional and national biosafety committees, regulatory officials, and ad hoc reviewers a thorough grounding in how to apply science to the risk assessment/ risk management process. It affords practical experience in reviewing field-test proposals in accordance with national guidelines. In selecting candidates for technical biosafety training, managers must make the difficult choice between exposing many people to at least one biosafety course versus providing recurrent training to fewer people. There is value in providing initial training for as many people as resources will allow, as it tends to create the critical massî needed to keep biosafety under discussion once the workshop is over. Further, it is only realistic to acknowledge that conflicts of interest can arise when a shortage of technically trained people results in the same few individuals being involved with both the research and the review.

On the other hand, becoming competent and having a sense of confidence in making biosafety decisions comes with practice. Providing more training to fewer people means that when there are field-test applications to evaluate, those who have had opportunities to review a variety of proposals under the guidance of experienced instructors will bring more skill and confidence to the review committee. In selecting candidates to serve on biosafety review committees, managers need to identify people with suitable expertise who can bring good analytical skills to the job. The ability to discriminate, to discern what is relevant and what is tangential; the capacity to recognize what is scientifically credible and what is merely anecdotal; the competence to judge the applicability and acceptability of data; all of these skills are crucial to making sound biosafety decisions.

It takes time and money to establish and maintain a biosafety system. A partial list of commonly encountered costs includes the following:

— *Meetings*: Biosafety review committees and teams that draft guidelines are usually composed of people from different institutions, often in different cities.

— *Consultations with international experts*: Outside expertise used in the early stages allows managers to capitalize on what has been learned elsewhere.

— *Access to information*: People in the biosafety system and outside stakeholders need to be informed of national and international biosafety information.

— *Monitoring*: Costs depend on who follows up, what they look for, and what they report.

— *Record keeping/information management*: Paperwork needs to be centrally organized and accessible. Documents and procedures need to be distributed to interested parties.

— *Administrative overhead*: Telephone, fax, photocopying, e-mail, and Internet access charges are recurring expenses.

Anticipating these costs, and knowing that there will likely be others, gives managers a more realistic basis for establishing a biosafety implementation plan.

Managing Biosafety

An effective biosafety system fosters the judicious use of biotechnology to improve agricultural productivity and food quality and secure economic benefits, while protecting human health and the environment. Such a system is one in which the four elements are balanced and effective:

- The guidelines clearly define the structure of the biosafety system, the roles and responsibilities of those involved, and how the review process is to operate.
- The people are knowledgeable and well-trained, confident in their ability to make decisions, and supported by their institutions.
- The review process is based on up-to-date scientific information; focuses on specific combinations of crop, gene, and environment; promotes appropriate risk management practices; and balances risks against benefits;
- Feedback mechanisms are used to incorporate new information and revise the system as needed.

Criteria such as these are open to interpretation and not always measurable. More objective measures of success can provide a better assessment of how well the biosafety system is working. Objective criteria to use could include the following:

- The system is operational: applications are being submitted, reviews are being conducted, and decisions are being made.
- Decisions have substance and weight; there is no need to seek repeated consensus on the same issues.
- Proponents, reviewers, and regulators have participated in appropriate training courses; training is an ongoing effort.

— Compliance is the norm; enforcement measures are seldom if ever needed.

— Stakeholders remain informed and involved by attending regular update and advisory meetings.

— Stakeholders collectively view the system as adequate; a consistent message is communicated to the public.

9

Cartagena Protocol on Biosafety to the Convention on Biological Diversity

The Parties to this Protocol,

Being Parties to the Convention on Biological Diversity, hereinafter referred to as "the Convention",

Recalling Article 19, paragraphs 3 and 4, and Articles 8 (g) and 17 of the Convention,

Recalling also decision II/5 of 17 November 1995 of the Conference of the Parties to the Convention to develop a Protocol on biosafety, specifically focusing on transboundary movement of any living modified organism resulting from modern biotechnology that may have adverse effect on the conservation and sustainable use of biological diversity, setting out for consideration, in particular, appropriate procedures for advance informed agreement,

Reaffirming the precautionary approach contained in Principle 15 of the Rio Declaration on Environment and Development,

Aware of the rapid expansion of modern biotechnology and the growing public concern over its potential adverse effects on biological diversity, taking also into account risks to human health,

Recognizing that modern biotechnology has great potential for human well-being if developed and used with adequate safety measures for the environment and human health,

Recognizing also the crucial importance to humankind of centres of origin and centres of genetic diversity,

Taking into account the limited capabilities of many countries, particularly developing countries, to cope with the nature and scale of known and potential risks associated with living modified organisms,

Recognizing that trade and environment agreements should be mutually supportive with a view to achieving sustainable development,

Emphasizing that this Protocol shall not be interpreted as implying a change in the rights and obligations of a Party under any existing international agreements,

Understanding that the above recital is not intended to subordinate this Protocol to other international agreements, Have agreed as follows:

Article 1
Objectives

In accordance with the precautionary approach contained in Principle 15 of the Rio Declaration on Environment and Development, the objective of this Protocol is to contribute to ensuring an adequate level of protection in the field of the safe transfer, handling and use of living modified organisms resulting from modern biotechnology that may have adverse effects on the conservation and sustainable use of biological diversity, taking also into account risks to human health, and specifically focusing on transboundary movements.

Article 2
General Provisions

1. Each Party shall take necessary and appropriate legal, administrative and other measures to implement its obligations under this Protocol.
2. The Parties shall ensure that the development, handling, transport, use, transfer and release of any living modified organisms are undertaken in a manner that prevents or reduces the risks to biological diversity, taking also into account risks to human health.
3. Nothing in this Protocol shall affect in any way the sovereignty of States over their territorial sea established in accordance with international law, and the sovereign rights and the jurisdiction which States have in their exclusive economic zones and their continental shelves in accordance with international law, and the exercise by ships and aircraft of all States of navigational rights and freedoms as provided for in international law and as reflected in relevant international instruments.
4. Nothing in this Protocol shall be interpreted as restricting the right of a Party to take action that is more protective of the conservation and sustainable use of biological diversity than that called for in this Protocol, provided that such action is consistent with the objective and the provisions of this Protocol and is in accordance with that Party's other obligations under international law.
5. The Parties are encouraged to take into account, as appropriate, available expertise, instruments and work undertaken in international forums with competence in the area of risks to human health.

Article 3
Use of Terms

For the purposes of this Protocol:

(a) "Conference of the Parties" means the Conference of the Parties to the Convention;

(b) "Contained use" means any operation, undertaken within a facility, installation or other physical structure, which involves living modified organisms that are controlled by specific measures that effectively limit their contact with, and their impact on, the external environment;

(c) "Export" means intentional transboundary movement from one Party to another Party;

(d) "Exporter" means any legal or natural person, under the jurisdiction of the Party of export, who arranges for a living modified organism to be exported;

(e) "Import" means intentional transboundary movement into one Party from another Party;

(f) "Importer" means any legal or natural person, under the jurisdiction of the Party of import, who arranges for a living modified organism to be imported;

(g) "Living modified organism" means any living organism that possesses a novel combination of genetic material obtained through the use of modern biotechnology;

(h) "Living organism" means any biological entity capable of transferring or replicating genetic material, including sterile organisms, viruses and viroids;

(i) "Modern biotechnology" means the application of:

   a. In vitro nucleic acid techniques, including recombinant deoxyribonucleic acid (DNA) and direct injection of nucleic acid into cells or organelles, or

   b. Fusion of cells beyond the taxonomic family, that overcome natural physiological reproductive or recombination barriers and that are not techniques used in traditional breeding and selection;

(j) "Regional economic integration organization" means an organization constituted by sovereign States of a given region, to which its member States have transferred competence in respect of matters governed by this Protocol and which has been duly authorized, in accordance with its internal procedures, to sign, ratify, accept, approve or accede to it;

(k) "Transboundary movement" means the movement of a living modified organism from one Party to another Party, save that for the purposes of Articles 17 and 24 transboundary movement extends to movement between Parties and non-Parties.

Article 4
Scope

This Protocol shall apply to the transboundary movement, transit, handling and use of all living modified organisms that may have adverse effects on the conservation and sustainable use of biological diversity, taking also into account risks to human health.

Article 5
Pharmaceuticals

Notwithstanding Article 4 and without prejudice to any right of a Party to subject all living modified organisms to risk assessment prior to the making of decisions on import, this Protocol shall not apply to the transboundary movement of living modified organisms which are pharmaceuticals for humans that are addressed by other relevant international agreements or organisations.

Article 6
Transit and Contained Use

1. Notwithstanding Article 4 and without prejudice to any

right of a Party of transit to regulate the transport of living modified organisms through its territory and make available to the Biosafety Clearing-House, any decision of that Party, subject to Article 2, paragraph 3, regarding the transit through its territory of a specific living modified organism, the provisions of this Protocol with respect to the advance informed agreement procedure shall not apply to living modified organisms in transit.

2. Notwithstanding Article 4 and without prejudice to any right of a Party to subject all living modified organisms to risk assessment prior to decisions on import and to set standards for contained use within its jurisdiction, the provisions of this Protocol with respect to the advance informed agreement procedure shall not apply to the transboundary movement of living modified organisms destined for contained use undertaken in accordance with the standards of the Party of import.

Article 7
Application of the Advance Informed Agreement Procedure

1. Subject to Articles 5 and 6, the advance informed agreement procedure in Articles 8 to 10 and 12 shall apply prior to the first intentional transboundary movement of living modified organisms for intentional introduction into the environment of the Party of import.

2. "Intentional introduction into the environment" in paragraph 1 above, does not refer to living modified organisms intended for direct use as food or feed, or for processing.

3. Article 11 shall apply prior to the first transboundary movement of living modified organisms intended for direct use as food or feed, or for processing.

4. The advance informed agreement procedure shall not apply to the intentional transboundary movement of living modified organisms identified in a decision of the Conference of the Parties serving as the meeting of the Parties to this Protocol as being not likely to have adverse effects on the conservation and sustainable use of biological diversity, taking also into account risks to human health.

Article 8
Notification

1. The Party of export shall notify, or require the exporter to ensure notification to, in writing, the competent national authority of the Party of import prior to the intentional transboundary movement of a living modified organism that falls within the scope of Article 7, paragraph 1.

2. The Party of export shall ensure that there is a legal requirement for the accuracy of information provided by the exporter.

Article 9
Acknowldgement of Receipt of Notification

1. The Party of import shall acknowledge receipt of the notification, in writing, to the notifier within ninety days of its receipt.

2. The acknowledgement shall state:

   (a) The date of receipt of the notification;

   (b) Whether the notification, prima facie, contains the information referred to in Article 8;

   (c) Whether to proceed according to the domestic regulatory framework of the Party of import or according to the procedure specified in Article 10.

3. The domestic regulatory framework referred to in paragraph 2 (c) above, shall be consistent with this Protocol.

4. A failure by the Party of import to acknowledge receipt of a notification shall not imply its consent to an intentional transboundary movement.

Article 10
Decision Procedure

1. Decisions taken by the Party of import shall be in accordance with Article 15.

2. The Party of import shall, within the period of time referred to in Article 9, inform the notifier, in writing, whether the intentional transboundary movement may proceed:

   (a) Only after the Party of import has given its written consent; or

   (b) After no less than ninety days without a subsequent written consent.

3. Within two hundred and seventy days of the date of receipt of notification, the Party of import shall communicate, in writing, to the notifier and to the Biosafety Clearing-House the decision referred to in paragraph 2 (a) above:

   (a) Approving the import, with or without conditions, including how the decision will apply to subsequent imports of the same living modified organism;

   (b) Prohibiting the import;

   (c) Requesting additional relevant information in accordance with its domestic regulatory framework or Annex I; in calculating the time within which the Party of import is to respond, the number of days it has to wait for additional relevant information shall not be taken into account; or

(d) Informing the notifier that the period specified in this paragraph is extended by a defined period of time.

4. Except in a case in which consent is unconditional, a decision under paragraph 3 above, shall set out the reasons on which it is based.

5. A failure by the Party of import to communicate its decision within two hundred and seventy days of the date of receipt of the notification shall not imply its consent to an intentional transboundary movement.

6. Lack of scientific certainty due to insufficient relevant scientific information and knowledge regarding the extent of the potential adverse effects of a living modified organism on the conservation and sustainable use of biological diversity in the Party of import, taking also into account risks to human health, shall not prevent that Party from taking a decision, as appropriate, with regard to the import of the living modified organism in question as referred to in paragraph 3 above, in order to avoid or minimize such potential adverse effects.

7. The Conference of the Parties serving as the meeting of the Parties shall, at its first meeting, decide upon appropriate procedures and mechanisms to facilitate decision-making by Parties of import.

Article 11
Procedure for Living Modified Organisms Intended for Direct Use as Food or Feed, or for Processinng

1. A Party that makes a final decision regarding domestic use, including placing on the market, of a living modified organism that may be subject to transboundary movement for direct use as food or feed, or for processing

shall, within fifteen days of making that decision, inform the Parties through the Biosafety Clearing-House. This information shall contain, at a minimum, the information specified in Annex II. The Party shall provide a copy of the information, in writing, to the national focal point of each Party that informs the Secretariat in advance that it does not have access to the Biosafety Clearing-House. This provision shall not apply to decisions regarding field trials.

2. The Party making a decision under paragraph 1 above, shall ensure that there is a legal requirement for the accuracy of information provided by the applicant.

3. Any Party may request additional information from the authority identified in paragraph (b) of Annex II.

4. A Party may take a decision on the import of living modified organisms intended for direct use as food or feed, or for processing, under its domestic regulatory framework that is consistent with the objective of this Protocol.

5. Each Party shall make available to the Biosafety Clearing-House copies of any national laws, regulations and guidelines applicable to the import of living modified organisms intended for direct use as food or feed, or for processing, if available.

6. A developing country Party or a Party with an economy in transition may, in the absence of the domestic regulatory framework referred to in paragraph 4 above, and in exercise of its domestic jurisdiction, declare through the Biosafety Clearing-House that its decision prior to the first import of a living modified organism intended for direct use as food or feed, or for processing, on which information has been provided under paragraph 1 above, will be taken according to the following:

(a) A risk assessment undertaken in accordance with Article 15; and

(b) A decision made within a predictable timeframe, not exceeding two hundred and seventy days.

7. Failure by a Party to communicate its decision according to paragraph 6 above, shall not imply its consent or refusal to the import of a living modified organism intended for direct use as food or feed, or for processing, unless otherwise specified by the Party.

8. Lack of scientific certainty due to insufficient relevant scientific information and knowledge regarding the extent of the potential adverse effects of a living modified organism on the conservation and sustainable use of biological diversity in the Party of import, taking also into account risks to human health, shall not prevent that Party from taking a decision, as appropriate, with regard to the import of that living modified organism intended for direct use as food or feed, or for processing, in order to avoid or minimize such potential adverse effects.

9. A Party may indicate its needs for financial and technical assistance and capacity-building with respect to living modified organisms intended for direct use as food or feed, or for processing. Parties shall cooperate to meet these needs in accordance with Articles 22 and 28.

Article 12
Review of Decisions

1. A Party of import may, at any time, in light of new scientific information on potential adverse effects on the conservation and sustainable use of biological diversity, taking also into account the risks to human health, review and change a decision regarding an intentional transboundary movement. In such case, the Party shall, within thirty days, inform any notifier that has previously

notified movements of the living modified organism referred to in such decision, as well as the Biosafety Clearing-House, and shall set out the reasons for its decision.

2. A Party of export or a notifier may request the Party of import to review a decision it has made in respect of it under Article 10 where the Party of export or the notifier considers that:

   (a) A change in circumstances has occurred that may influence the outcome of the risk assessment upon which the decision was based; or

   (b) Additional relevant scientific or technical information has become available.

3. The Party of import shall respond in writing to such a request within ninety days and set out the reasons for its decision.

4. The Party of import may, at its discretion, require a risk assessment for subsequent imports.

Article 13
Simplified Procedure

1. A Party of import may, provided that adequate measures are applied to ensure the safe intentional transboundary movement of living modified organisms in accordance with the objective of this Protocol, specify in advance to the Biosafety Clearing-House:

   (a) Cases in which intentional transboundary movement to it may take place at the same time as the movement is notified to the Party of import; and

   (b) Imports of living modified organisms to it to be exempted from the advance informed agreement procedure.

Notifications under subparagraph (a) above, may apply to subsequent similar movements to the same Party.

2. The information relating to an intentional transboundary movement that is to be provided in the notifications referred to in paragraph 1

(a) above, shall be the information specified in Annex I.

Article 14
Bilateral, Regional and Multilateral Agreements and Arrangements

1. Parties may enter into bilateral, regional and multilateral agreements and arrangements regarding intentional transboundary movements of living modified organisms, consistent with the objective of this Protocol and provided that such agreements and arrangements do not result in a lower level of protection than that provided for by the Protocol.

2. The Parties shall inform each other, through the Biosafety Clearing-House, of any such bilateral, regional and multilateral agreements and arrangements that they have entered into before or after the date of entry into force of this Protocol.

3. The provisions of this Protocol shall not affect intentional transboundary movements that take place pursuant to such agreements and arrangements as between the parties to those agreements or arrangements.

4. Any Party may determine that its domestic regulations shall apply with respect to specific imports to it and shall notify the Biosafety Clearing-House of its decision.

Article 15
Risk Assessment

1. Risk assessments undertaken pursuant to this Protocol

shall be carried out in a scientifically sound manner, in accordance with Annex III and taking into account recognized risk assessment techniques. Such risk assessments shall be based, at a minimum, on information provided in accordance with Article 8 and other available scientific evidence in order to identify and evaluate the possible adverse effects of living modified organisms on the conservation and sustainable use of biological diversity, taking also into account risks to human health.

2. The Party of import shall ensure that risk assessments are carried out for decisions taken under Article 10. It may require the exporter to carry out the risk assessment.

3. The cost of risk assessment shall be borne by the notifier if the Party of import so requires.

Article 16
Risk Management

1. The Parties shall, taking into account Article 8 (g) of the Convention, establish and maintain appropriate mechanisms, measures and strategies to regulate, manage and control risks identified in the risk assessment provisions of this Protocol associated with the use, handling and transboundary movement of living modified organisms.

2. Measures based on risk assessment shall be imposed to the extent necessary to prevent adverse effects of the living modified organism on the conservation and sustainable use of biological diversity, taking also into account risks to human health, within the territory of the Party of import.

3. Each Party shall take appropriate measures to prevent unintentional transboundary movements of living modified organisms, including such measures as

requiring a risk assessment to be carried out prior to the first release of a living modified organism.

4. Without prejudice to paragraph 2 above, each Party shall endeavour to ensure that any living modified organism, whether imported or locally developed, has undergone an appropriate period of observation that is commensurate with its life-cycle or generation time before it is put to its intended use.

5. Parties shall cooperate with a view to:

   (a) Identifying living modified organisms or specific traits of living modified organisms that may have adverse effects on the conservation and sustainable use of biological diversity, taking also into account risks to human health; and

   (b) Taking appropriate measures regarding the treatment of such living modified organisms or specific traits.

Article 17
Unintentional Transboundary Movements and Emergency Measures

1. Each Party shall take appropriate measures to notify affected or potentially affected States, the Biosafety Clearing-House and, where appropriate, relevant international organizations, when it knows of an occurrence under its jurisdiction resulting in a release that leads, or may lead, to an unintentional transboundary movement of a living modified organism that is likely to have significant adverse effects on the conservation and sustainable use of biological diversity, taking also into account risks to human health in such States. The notification shall be provided as soon as the Party knows of the above situation.

2. Each Party shall, no later than the date of entry into force of this Protocol for it, make available to the Biosafety

Clearing-House the relevant details setting out its point of contact for the purposes of receiving notifications under this Article.

3. Any notification arising from paragraph 1 above, should include:

(a) Available relevant information on the estimated quantities and relevant characteristics and/or traits of the living modified organism;

(b) Information on the circumstances and estimated date of the release, and on the use of the living modified organism in the originating Party;

(c) Any available information about the possible adverse effects on the conservation and sustainable use of biological diversity, taking also into account risks to human health, as well as available information about possible risk management measures;

(d) Any other relevant information; and

(e) A point of contact for further information.

4. In order to minimize any significant adverse effects on the conservation and sustainable use of biological diversity, taking also into account risks to human health, each Party, under whose jurisdiction the release of the living modified organism referred to in paragraph 1 above, occurs, shall immediately consult the affected or potentially affected States to enable them to determine appropriate responses and initiate necessary action, including emergency measures.

Article 18
Handling, Transport, Packaging and Identification

1. In order to avoid adverse effects on the conservation and sustainable use of biological diversity, taking also into account risks to human health, each Party shall take

necessary measures to require that living modified organisms that are subject to intentional transboundary movement within the scope of this Protocol are handled, packaged and transported under conditions of safety, taking into consideration relevant international rules and standards.

2. Each Party shall take measures to require that documentation accompanying:

   (a) Living modified organisms that are intended for direct use as food or feed, or for processing, clearly identifies that they "may contain" living modified organisms and are not intended for intentional introduction into the environment, as well as a contact point for further information. The Conference of the Parties serving as the meeting of the Parties to this Protocol shall take a decision on the detailed requirements for this purpose, including specification of their identity and any unique identification, no later than two years after the date of entry into force of this Protocol;

   (b) Living modified organisms that are destined for contained use clearly identifies them as living modified organisms; and specifies any requirements for the safe handling, storage, transport and use, the contact point for further information, including the name and address of the individual and institution to whom the living modified organisms are consigned; and

   (c) Living modified organisms that are intended for intentional introduction into the environment of the Party of import and any other living modified organisms within the scope of the Protocol, clearly identifies them as living modified organisms; specifies the identity and relevant traits and/or characteristics, any requirements for the safe handling, storage, transport and use, the contact point for further information and, as appropriate, the name and address of the importer and exporter; and contains a declaration that the movement is in conformity with the requirements of this Protocol applicable to the exporter.

3. The Conference of the Parties serving as the meeting of the Parties to this Protocol shall consider the need for and modalities of developing standards with regard to identification, handling, packaging and transport practices, in consultation with other relevant international bodies.

Article 19
Competent National Authorities and National Focal Points

1. Each Party shall designate one national focal point to be responsible on its behalf for liaison with the Secretariat. Each Party shall also designate one or more competent national authorities, which shall be responsible for performing the administrative functions required by this Protocol and which shall be authorized to act on its behalf with respect to those functions. A Party may designate a single entity to fulfil the functions of both focal point and competent national authority.

2. Each Party shall, no later than the date of entry into force of this Protocol for it, notify the Secretariat of the names and addresses of its focal point and its competent national authority or authorities. Where a Party designates more than one competent national authority, it shall convey to the Secretariat, with its notification thereof, relevant information on the respective responsibilities of those authorities. Where applicable, such information shall, at a minimum, specify which competent authority is responsible for which type of living modified organism. Each Party shall forthwith notify the Secretariat of any

changes in the designation of its national focal point or in the name and address or responsibilities of its competent national authority or authorities.

3. The Secretariat shall forthwith inform the Parties of the notifications it receives under paragraph 2 above, and shall also make such information available through the Biosafety Clearing-House.

Article 20
Information Sharing and the Biosafety Clearing-house

1. A Biosafety Clearing-House is hereby established as part of the clearing-house mechanism under Article 18, paragraph 3, of the Convention, in order to:

   (a) Facilitate the exchange of scientific, technical, environmental and legal information on, and experience with, living modified organisms; and

   (b) Assist Parties to implement the Protocol, taking into account the special needs of developing country Parties, in particular the least developed and small island developing States among them, and countries with economies in transition as well as countries that are centres of origin and centres of genetic diversity.

2. The Biosafety Clearing-House shall serve as a means through which information is made available for the purposes of paragraph 1 above. It shall provide access to information made available by the Parties relevant to the implementation of the Protocol. It shall also provide access, where possible, to other international biosafety information exchange mechanisms.

3. Without prejudice to the protection of confidential information, each Party shall make available to the Biosafety Clearing-House any information required to be made available to the Biosafety Clearing-House under this Protocol, and:

(a) Any existing laws, regulations and guidelines for implementation of the Protocol, as well as information required by the Parties for the advance informed agreement procedure;

(b) Any bilateral, regional and multilateral agreements and arrangements;

(c) Summaries of its risk assessments or environmental reviews of living modified organisms generated by its regulatory process, and carried out in accordance with Article 15, including, where appropriate, relevant information regarding products thereof, namely, processed materials that are of living modified organism origin, containing detectable novel combinations of replicable genetic material obtained through the use of modern biotechnology;

(d) Its final decisions regarding the importation or release of living modified organisms; and

(e) Reports submitted by it pursuant to Article 33, including those on implementation of the advance informed agreement procedure.

4. The modalities of the operation of the Biosafety Clearing-House, including reports on its activities, shall be considered and decided upon by the Conference of the Parties serving as the meeting of the Parties to this Protocol at its first meeting, and kept under review thereafter.

Article 21
Confidential Information

1. The Party of import shall permit the notifier to identify information submitted under the procedures of this Protocol or required by the Party of import as part of the advance informed agreement procedure of the Protocol

that is to be treated as confidential. Justification shall be given in such cases upon request.

2. The Party of import shall consult the notifier if it decides that information identified by the notifier as confidential does not qualify for such treatment and shall, prior to any disclosure, inform the notifier of its decision, providing reasons on request, as well as an opportunity for consultation and for an internal review of the decision prior to disclosure.

3. Each Party shall protect confidential information received under this Protocol, including any confidential information received in the context of the advance informed agreement procedure of the Protocol. Each Party shall ensure that it has procedures to protect such information and shall protect the confidentiality of such information in a manner no less favourable than its treatment of confidential information in connection with domestically produced living modified organisms.

4. The Party of import shall not use such information for a commercial purpose, except with the written consent of the notifier.

5. If a notifier withdraws or has withdrawn a notification, the Party of import shall respect the confidentiality of commercial and industrial information, including research and development information as well as information on which the Party and the notifier disagree as to its confidentiality.

6. Without prejudice to paragraph 5 above, the following information shall not be considered confidential:

    (a) The name and address of the notifier;

    (b) A general description of the living modified organism or organisms;

(c) A summary of the risk assessment of the effects on the conservation and sustainable use of biological diversity, taking also into account risks to human health; and

(d) Any methods and plans for emergency response.

Article 22
Capacity-building

1. The Parties shall cooperate in the development and/or strengthening of human resources and institutional capacities in biosafety, including biotechnology to the extent that it is required for biosafety, for the purpose of the effective implementation of this Protocol, in developing country Parties, in particular the least developed and small island developing States among them, and in Parties with economies in transition, including through existing global, regional, subregional and national institutions and organizations and, as appropriate, through facilitating private sector involvement.

2. For the purposes of implementing paragraph 1 above, in relation to cooperation, the needs of developing country Parties, in particular the least developed and small island developing States among them, for financial resources and access to and transfer of technology and knowhow in accordance with the relevant provisions of the Convention, shall be taken fully into account for capacity-building in biosafety. Cooperation in capacity-building shall, subject to the different situation, capabilities and requirements of each Party, include scientific and technical training in the proper and safe management of biotechnology, and in the use of risk assessment and risk management for biosafety, and the enhancement of technological and institutional capacities in biosafety. The needs of Parties with economies in transition shall also

be taken fully into account for such capacitybuilding in biosafety.

Article 23
Public Awareness and Participation

1. The Parties shall:

   (a) Promote and facilitate public awareness, education and participation concerning the safe transfer, handling and use of living modified organisms in relation to the conservation and sustainable use of biological diversity, taking also into account risks to human health. In doing so, the Parties shall cooperate, as appropriate, with other States and international bodies;

   (b) Endeavour to ensure that public awareness and education encompass access to information on living modified organisms identified in accordance with this Protocol that may be imported.

2. The Parties shall, in accordance with their respective laws and regulations, consult the public in the decision-making process regarding living modified organisms and shall make the results of such decisions available to the public, while respecting confidential information in accordance with Article 21.

3. Each Party shall endeavour to inform its public about the means of public access to the Biosafety Clearing-House.

Article 24
Non-parties

1. Transboundary movements of living modified organisms between Parties and non-Parties shall be consistent with the objective of this Protocol. The Parties may enter into bilateral, regional and multilateral agreements and

arrangements with non-Parties regarding such transboundary movements.

2. The Parties shall encourage non-Parties to adhere to this Protocol and to contribute appropriate information to the Biosafety Clearing-House on living modified organisms released in, or moved into or out of, areas within their national jurisdictions.

Article 25
Illegal Transboundary Movements

1. Each Party shall adopt appropriate domestic measures aimed at preventing and, if appropriate, penalizing transboundary movements of living modified organisms carried out in contravention of its domestic measures to implement this Protocol. Such movements shall be deemed illegal transboundary movements.

2. In the case of an illegal transboundary movement, the affected Party may request the Party of origin to dispose, at its own expense, of the living modified organism in question by repatriation or destruction, as appropriate.

3. Each Party shall make available to the Biosafety Clearing-House information concerning cases of illegal transboundary movements pertaining to it.

Article 26
Socio-economic Considerations

1. The Parties, in reaching a decision on import under this Protocol or under its domestic measures implementing the Protocol, may take into account, consistent with their international obligations, socioeconomic considerations arising from the impact of living modified organisms on the conservation and sustainable use of biological diversity, especially with regard to the value of biological diversity to indigenous and local communities.

2. The Parties are encouraged to cooperate on research and information exchange on any socio-economic impacts of living modified organisms, especially on indigenous and local communities.

Article 27
Liability and Redress

The Conference of the Parties serving as the meeting of the Parties to this Protocol shall, at its first meeting, adopt a process with respect to the appropriate elaboration of international rules and procedures in the field of liability and redress for damage resulting from transboundary movements of living modified organisms, analysing and taking due account of the ongoing processes in international law on these matters, and shall endeavour to complete this process within four years.

Article 28
Financial Mechanism and Resources

1. In considering financial resources for the implementation of this Protocol, the Parties shall take into account the provisions of Article 20 of the Convention.

2. The financial mechanism established in Article 21 of the Convention shall, through the institutional structure entrusted with its operation, be the financial mechanism for this Protocol.

3. Regarding the capacity-building referred to in Article 22 of this Protocol, the Conference of the Parties serving as the meeting of the Parties to this Protocol, in providing guidance with respect to the financial mechanism referred to in paragraph 2 above, for consideration by the Conference of the Parties, shall take into account the need for financial resources by developing country Parties, in particular the least developed and the small island developing States among them.

4. In the context of paragraph 1 above, the Parties shall also take into account the needs of the developing country Parties, in particular the least developed and the small island developing States among them, and of the Parties with economies in transition, in their efforts to identify and implement their capacity-building requirements for the purposes of the implementation of this Protocol.

5. The guidance to the financial mechanism of the Convention in relevant decisions of the Conference of the Parties, including those agreed before the adoption of this Protocol, shall apply, mutatis mutandis, to the provisions of this Article.

6. The developed country Parties may also provide, and the developing country Parties and the Parties with economies in transition avail themselves of, financial and technological resources for the implementation of the provisions of this Protocol through bilateral, regional and multilateral channels.

Article 29
Conference of the Parties Serving as the Meeting of the Parties to this Protocol

1. The Conference of the Parties shall serve as the meeting of the Parties to this Protocol.

2. Parties to the Convention that are not Parties to this Protocol may participate as observers in the proceedings of any meeting of the Conference of the Parties serving as the meeting of the Parties to this Protocol. When the Conference of the Parties serves as the meeting of the Parties to this Protocol, decisions under this Protocol shall be taken only by those that are Parties to it.

3. When the Conference of the Parties serves as the meeting of the Parties to this Protocol, any member of the bureau of the Conference of the Parties representing a Party to

the Convention but, at that time, not a Party to this Protocol, shall be substituted by a member to be elected by and from among the Parties to this Protocol.

4. The Conference of the Parties serving as the meeting of the Parties to this Protocol shall keep under regular review the implementation of this Protocol and shall make, within its mandate, the decisions necessary to promote its effective implementation. It shall perform the functions assigned to it by this Protocol and shall:

(a) Make recommendations on any matters necessary for the implementation of this Protocol;

(b) Establish such subsidiary bodies as are deemed necessary for the implementation of this Protocol;

(c) Seek and utilize, where appropriate, the services and cooperation of, and information provided by, competent international organizations and intergovernmental and non-governmental bodies;

(d) Establish the form and the intervals for transmitting the information to be submitted in accordance with Article 33 of this Protocol and consider such information as well as reports submitted by any subsidiary body;

(e) Consider and adopt, as required, amendments to this Protocol and its annexes, as well as any additional annexes to this Protocol, that are deemed necessary for the implementation of this Protocol; and

(f) Exercise such other functions as may be required for the implementation of this Protocol.

5. The rules of procedure of the Conference of the Parties and financial rules of the Convention shall be applied, mutatis mutandis, under this Protocol, except as may be otherwise decided by consensus by the Conference of the Parties serving as the meeting of the Parties to this Protocol.

6. The first meeting of the Conference of the Parties serving as the meeting of the Parties to this Protocol shall be convened by the Secretariat in conjunction with the first meeting of the Conference of the Parties that is scheduled after the date of the entry into force of this Protocol. Subsequent ordinary meetings of the Conference of the Parties serving as the meeting of the Parties to this Protocol shall be held in conjunction with ordinary meetings of the Conference of the Parties, unless otherwise decided by the Conference of the Parties serving as the meeting of the Parties to this Protocol.

7. Extraordinary meetings of the Conference of the Parties serving as the meeting of the Parties to this Protocol shall be held at such other times as may be deemed necessary by the Conference of the Parties serving as the meeting of the Parties to this Protocol, or at the written request of any Party, provided that, within six months of the request being communicated to the Parties by the Secretariat, it is supported by at least one third of the Parties.

8. The United Nations, its specialized agencies and the International Atomic Energy Agency, as well as any State member thereof or observers thereto not party to the Convention, may be represented as observers at meetings of the Conference of the Parties serving as the meeting of the Parties to this Protocol. Any body or agency, whether national or international, governmental or nongovernmental, that is qualified in matters covered by this Protocol and that has informed the Secretariat of its wish to be represented at a meeting of the Conference of the Parties serving as a meeting of the Parties to this Protocol as an observer, may be so admitted, unless at least one third of the Parties present object. Except as otherwise provided in this Article, the admission and participation of observers shall be subject to the rules of procedure, as referred to in paragraph 5 above.

Article 30
Subsidiary Bodies

1. Any subsidiary body established by or under the Convention may, upon a decision by the Conference of the Parties serving as the meeting of the Parties to this Protocol, serve the Protocol, in which case the meeting of the Parties shall specify which functions that body shall exercise.

2. Parties to the Convention that are not Parties to this Protocol may participate as observers in the proceedings of any meeting of any such subsidiary bodies. When a subsidiary body of the Convention serves as a subsidiary body to this Protocol, decisions under the Protocol shall be taken only by the Parties to the Protocol.

3. When a subsidiary body of the Convention exercises its functions with regard to matters concerning this Protocol, any member of the bureau of that subsidiary body representing a Party to the Convention but, at that time, not a Party to the Protocol, shall be substituted by a member to be elected by and from among the Parties to the Protocol.

Article 31
Secretariat

1. The Secretariat established by Article 24 of the Convention shall serve as the secretariat to this Protocol.

2. Article 24, paragraph 1, of the Convention on the functions of the Secretariat shall apply, mutatis mutandis, to this Protocol.

3. To the extent that they are distinct, the costs of the secretariat services for this Protocol shall be met by the Parties hereto. The Conference of the Parties serving as the meeting of the Parties to this Protocol shall, at its first

meeting, decide on the necessary budgetary arrangements to this end.

Article 32
Relationship with the Convention

Except as otherwise provided in this Protocol, the provisions of the Convention relating to its protocols shall apply to this Protocol.

Article 33
Monitoring and Reporting

Each Party shall monitor the implementation of its obligations under this Protocol, and shall, at intervals to be determined by the Conference of the Parties serving as the meeting of the Parties to this Protocol, report to the Conference of the Parties serving as the meeting of the Parties to this Protocol on measures that it has taken to implement the Protocol.

Article 34
Compliance

The Conference of the Parties serving as the meeting of the Parties to this Protocol shall, at its first meeting, consider and approve cooperative procedures and institutional mechanisms to promote compliance with the provisions of this Protocol and to address cases of non-compliance. These procedures and mechanisms shall include provisions to offer advice or assistance, where appropriate. They shall be separate from, and without prejudice to, the dispute settlement procedures and mechanisms established by Article 27 of the Convention.

Article 35
Assessment and Review

The Conference of the Parties serving as the meeting of the Parties to this Protocol shall undertake, five years after the entry into force of this Protocol and at least every five years thereafter, an evaluation of the effectiveness of the Protocol, including an assessment of its procedures and annexes.

Article 36
Signature

This Protocol shall be open for signature at the United Nations Office at Nairobi by States and regional economic integration organizations from 15 to 26 May 2000, and at United Nations Headquarters in New York from 5 June 2000 to 4 June 2001.

Article 37
Entry into Force

1. This Protocol shall enter into force on the ninetieth day after the date of deposit of the fiftieth instrument of ratification, acceptance, approval or accession by States or regional economic integration organizations that are Parties to the Convention.

2. This Protocol shall enter into force for a State or regional economic integration organization that ratifies, accepts or approves this Protocol or accedes thereto after its entry into force pursuant to paragraph 1 above, on the ninetieth day after the date on which that State or regional economic integration organization deposits its instrument of ratification, acceptance, approval or accession, or on the date on which the Convention enters into force for that State or regional economic integration organization, whichever shall be the later.

3. For the purposes of paragraphs 1 and 2 above, any instrument deposited by a regional economic integration organization shall not be counted as additional to those deposited by member States of such organization.

Article 38
Reservation

No reservations may be made to this Protocol.

Article 39
Withdrawal

1. At any time after two years from the date on which this Protocol has entered into force for a Party, that Party may withdraw from the Protocol by giving written notification to the Depositary.
2. Any such withdrawal shall take place upon expiry of one year after the date of its receipt by the Depositary, or on such later date as may be specified in the notification of the withdrawal.

Article 40
Authentic Texts

The original of this Protocol, of which the Arabic, Chinese, English, French, Russian and Spanish texts are equally authentic, shall be deposited with the Secretary-General of the United Nations. In witness whereof the undersigned, being duly authorized to that effect, have signed this Protocol.

Annex I
Information Required in Notifications Under Articles 8, 10 and 13

(a) Name, address and contact details of the exporter.

(b) Name, address and contact details of the importer.

(c) Name and identity of the living modified organism, as well as the domestic classification, if any, of the biosafety level of the living modified organism in the State of export.

(d) Intended date or dates of the transboundary movement, if known.

(e) Taxonomic status, common name, point of collection or acquisition, and characteristics of recipient organism or parental organisms related to biosafety.

(f) Centres of origin and centres of genetic diversity, if known, of the recipient organism and/or the parental organisms and a description of the habitats where the organisms may persist or proliferate.

(g) Taxonomic status, common name, point of collection or acquisition, and characteristics of the donor organism or organisms related to biosafety.

(h) Description of the nucleic acid or the modification introduced, the technique used, and the resulting characteristics of the living modified organism.

(i) Intended use of the living modified organism or products thereof, namely, processed materials that are of living modified organism origin, containing detectable novel combinations of replicable genetic material obtained through the use of modern biotechnology.

(j) Quantity or volume of the living modified organism to be transferred.

(k) A previous and existing risk assessment report consistent with Annex III.

(l) Suggested methods for the safe handling, storage, transport and use, including packaging, labelling, documentation, disposal and contingency procedures, where appropriate.

(m) Regulatory status of the living modified organism within the State of export (for example, whether it is prohibited in the State of export, whether there are other restrictions, or whether it has been approved for general release) and, if the living modified organism is banned in the State of export, the reason or reasons for the ban.

(n) Result and purpose of any notification by the exporter to other States regarding the living modified organism to be transferred.

(o) A declaration that the above-mentioned information is factually correct.

Annex II

Information Required Concerning Living Modified Organisms Intended for Direct Use as Food or Feed, or for Processing Under Article 11

(a) The name and contact details of the applicant for a decision for domestic use.

(b) The name and contact details of the authority responsible for the decision.

(c) Name and identity of the living modified organism.

(d) Description of the gene modification, the technique used, and the resulting characteristics of the living modified organism.

(e) Any unique identification of the living modified organism.

(f) Taxonomic status, common name, point of collection or acquisition, and characteristics of recipient organism or parental organisms related to biosafety.

(g) Centres of origin and centres of genetic diversity, if known, of the recipient organism and/or the parental organisms and a description of the habitats where the organisms may persist or proliferate.

(h) Taxonomic status, common name, point of collection or acquisition, and characteristics of the donor organism or organisms related to biosafety.

(i) Approved uses of the living modified organism.

(j) A risk assessment report consistent with Annex III.

(k) Suggested methods for the safe handling, storage, transport and use, including packaging, labelling, documentation, disposal and contingency procedures, where appropriate.

Annex III
Risk Assessment Under Article 15

Objective

1. The objective of risk assessment, under this Protocol, is to identify and evaluate the potential adverse effects of living modified organisms on the conservation and sustainable use of biological diversity in the likely potential receiving environment, taking also into account risks to human health.

Use of risk assessment

2. Risk assessment is, inter alia, used by competent authorities to make informed decisions regarding living modified organisms.

General principles

3. Risk assessment should be carried out in a scientifically sound and transparent manner, and can take into account expert advice of, and guidelines developed by, relevant international organizations.

4. Lack of scientific knowledge or scientific consensus should not necessarily be interpreted as indicating a particular level of risk, an absence of risk, or an acceptable risk.
5. Risks associated with living modified organisms or products thereof, namely, processed materials that are of living modified organism origin, containing detectable novel combinations of replicable genetic material obtained through the use of modern biotechnology, should be considered in the context of the risks posed by the non-modified recipients or parental organisms in the likely potential receiving environment.
6. Risk assessment should be carried out on a case-by-case basis.

The required information may vary in nature and level of detail from case to case, depending on the living modified organism concerned, its intended use and the likely potential receiving environment.

Methodology

7. The process of risk assessment may on the one hand give rise to a need for further information about specific subjects, which may be identified and requested during the assessment process, while on the other hand information on other subjects may not be relevant in some instances.
8. To fulfil its objective, risk assessment entails, as appropriate, the following steps:
   (a) An identification of any novel genotypic and phenotypic characteristics associated with the living modified organism that may have adverse effects on biological diversity in the likely potential receiving environment, taking also into account risks to human health;

(b) An evaluation of the likelihood of these adverse effects being realized, taking into account the level and kind of exposure of the likely potential receiving environment to the living modified organism;

(c) An evaluation of the consequences should these adverse effects be realized;

(d) An estimation of the overall risk posed by the living modified organism based on the evaluation of the likelihood and consequences of the identified adverse effects being realized;

(e) A recommendation as to whether or not the risks are acceptable or manageable, including, where necessary, identification of strategies to manage these risks; and

(f) Where there is uncertainty regarding the level of risk, it may be addressed by requesting further information on the specific issues of concern or by implementing appropriate risk management strategies and/or monitoring the living modified organism in the receiving environment.

Points to Consider

9. Depending on the case, risk assessment takes into account the relevant technical and scientific details regarding the characteristics of the following subjects:

(a) Recipient organism or parental organisms. The biological characteristics of the recipient organism or parental organisms, including information on taxonomic status, common name, origin, centres of origin and centres of genetic diversity, if known, and a description of the habitat where the organisms may persist or proliferate;

(b) Donor organism or organisms. Taxonomic status and common name, source, and the relevant biological characteristics of the donor organisms;

(c) Vector. Characteristics of the vector, including its identity, if any, and its source or origin, and its host range;

(d) Insert or inserts and/or characteristics of modification.

Genetic characteristics of the inserted nucleic acid and the function it specifies, and/or characteristics of the modification introduced;

(e) Living modified organism. Identity of the living modified organism, and the differences between the biological characteristics of the living modified organism and those of the recipient organism or parental organisms;

(f) Detection and identification of the living modified organism. Suggested detection and identification methods and their specificity, sensitivity and reliability;

(g) Information relating to the intended use. Information relating to the intended use of the living modified organism, including new or changed use compared to the recipient organism or parental organisms; and

(h) Receiving environment. Information on the location, geographical, climatic and ecological characteristics, including relevant information on biological diversity and centres of origin of the likely potential receiving environment.

Bibliography

Ernst & Young, On the threshold. The Asia Pacific *Perspective Global Biotechnology Report*, Ernst & Young, San Francisco, CA, USA, 2004.

Jayaraman, K.S. *Nature* 422, 6, 2003.

OECD, *An Introduction to the Biosafety Consensus Documents of OECD's Working Group on Harmonisation of Regulatory Oversight in Biotechnology*, OECD: Paris, 2005

James, C., "Global Status of Commercialized Biotech/GM Crops: 2004",ISAAA, 2004.

OECD, *Recommendation of the Council on the Use of Economic Instruments in Promoting the Conservation and Sustainable Use of Biodiversity*, OECD: Paris, 2004.

____________*Handbook of Market Creation for Biodiversity: Issues in Implementation*, OECD: Paris, 2004.

____________, *Harnessing Markets for Biodiversity: Towards Conservation and* Sustainable Use, 2003.

____________, *Perverse Incentives in Biodiversity Loss*. ENV/EPOC/GSP/BIO(2003)2/ FINAL, 2003.

____________, *Handbook of Biodiversity Valuation: A Guide for Policy-Makers*, OECD: Paris, 2002.

____________, *Valuation of Biodiversity Benefits: Selected Studies*, OECD: Paris, 2001.

____________, *Handbook of Incentive Measures for Biodiversity: Design and Implementation*, OECD: Paris, 1999.

Pandey, I.M. *Technovation* 18, 253–261, 1998.

Rao, S.R. *Asian Biotechnol. Dev. Rev.* 5, 1–14, 2003.

Index

Agricultural biotechnology 16, 41
Animal Toxicity Tests (ATT) 103

Biodiversity 1
Bio-engineered crops 16
Bio-engineering 63
Biological Weapons (BW) 148
Biosafety Clearing House (BCH) 9

Centres for Plant Molecular Biology (CPMB) 43
Chemical Weapons Convention (CWC) 160
Clearing-House Mechanism (CHM) 20
Clinical biotechnology 63, 113
Confidential business information 26
Contract Manufacturing Organizations (CMOs) 69
Contract Research Organizations (CROs) 69
Convention on Biological diversity (CBD) 20
Crop biotechnology 42

Deoxyribonucleic Acid (DNA) 234
Department of Biotechnology (DBT) 43
Diagnostic biotechnology 63
Domestic health biotechnology 53

Ecosystems 16
Environmental biotechnology 16
Ethnicnationalism 172

Food biotechnology 63
Foreign Investment Promotion Board (FIPB) 75

Genetic engineering (GE) 24
Genetically Engineered Crops 9
Genetically Engineered Organisms 197
Genetically modified (GM) 17
Genetically modified organisms (GMOs) 11
Global eco-tourism 3
Greenhouse gases 94

Human Clinical Trial (HCT) 104
Human Resource Development (HRD) 226

Indian Agricultural Research Institute (IARI) 45
Industrial biotechnology 63
Information management 226
Institutional Biosafety Committee (IBSC) 47

Intellectual Property (IP) 57
International Depository Authority (IDA) 97

Living Modified Organisms (LMOs) 19

Mass destruction 156
Medical biotechnology 63
Microbiology 77
Multilateral environmental agreements (MEAs) 40

Nano-biotechnology 63
Non-Governmental Organisations (NGOs) 24

Religious fanaticism 172

Risk management 216

Sound technology 27

Terminator technology 207
Terrorism 148
Therapeutic 63
Tissue culture 16
Transgenic crops 6

Venezuelan Equine Encephalomyelitis virus (VEE virus) 149

World Health Organization (WHO) 39, 199
World Trade Organisation (WTO) 26, 187